Anti-Capitalism

T0257153

Anti-Capitalism

A Marxist Introduction

Edited by
Alfredo Saad-Filho

Pluto Press
LONDON

First published 2003 by Pluto Press
345 Archway Road, London N6 5AA

www.plutobooks.com

British Library Cataloguing in Publication Data
A catalogue record for this book is available from the British Library

ISBN 978 0 7453 1893 6 paperback

Library of Congress Cataloging in Publication Data
applied for

10   9   8   7   6   5   4   3   2   1

Designed and produced for Pluto Press by
Chase Publishing Services, Fortescue, Sidmouth EX10 9QG
Typeset from disk by Stanford DTP Services, Towcester
Printed and bound in Great Britain by
Marston Book Services Limited, Didcot

Contents

Part III: Crisis and the Supercession of Capitalism

Acknowledgements

I am grateful to Andrew Brown, Sebastian Budgen, Ben Fine, John Holloway, Costas Lapavitsas, Simon Mohun, Pilgrim Tucker and Ellen Meiksins Wood for their support and assistance. Special thanks are due to Anne Beech of Pluto Press for her superb handling of this project.

Introduction[1]

Alfredo Saad-Filho

> The need of a constantly expanding market ... chases the bourgeoisie
> over the whole surface of the globe ... All old-established national
> industries ... are dislodged by new industries ... that no longer work up
> indigenous raw material, but raw material drawn from the remotest
> zones; industries whose products are consumed, not only at home, but
> in every quarter of the globe. In place of the old wants, satisfied by the
> productions of the country, we find new wants, requiring for their sat-
> isfaction the products of distant lands and climes ... The bourgeoisie,
> by the rapid improvement of all instruments of production, by the
> immensely facilitated means of communication, draws all ... nations
> into civilisation ... It compels all nations, on pain of extinction, to adopt
> the bourgeois mode of production; it compels them to introduce what it
> calls civilisation into their midst, i.e., to become bourgeois themselves.
> In one word, it creates a world after its own image.[2]

CAPITALISM AND ANTI-CAPITALISM

The *Communist Manifesto* rings even truer today than it did in 1848.
Key features of nineteenth-century capitalism are clearly recognis-
able, and even more strongly developed, in the early twenty-first
century. They include the internationalisation of trade, production
and finance, the growth of transnational corporations (TNCs), the
communications revolution, the diffusion of Western culture and
consumption patterns across the world, and so on.

Other traits of our age can also be found in the *Manifesto*. In the
early twenty-first century, powerful nations still rule the world by
political, economic and military means, and their gospel is zealously
preached by today's missionaries of neoliberalism. They follow in
the footsteps of their ancestors, who drew strength from the holy
trinity of Victorian imperialism: God, British capital and the Royal
Navy. Today's evangelists pay lip-service to human rights and the
elimination of poverty, but their faith lies elsewhere, in the sacred
tablets of copyright law and in the charter of the International
Monetary Fund (IMF). They travel to all corners of the globe and, in
spite of untold hardship in anonymous five-star hotels, tirelessly

preach submission to Wall Street and the US government. They will never take no for an answer. Native obduracy is initially explained away as ignorance or corruption, and then ridiculed. However, even saintly patience has its limits. Eventually, economic, diplomatic and other forms of pressure may become necessary. In extreme circumstances, the White House may be forced to bomb the enemy into submission, thus rendering another country safe for McDonald's.

It seems that, in spite of our fast cars, mobile phones and the internet, the world has not, after all, changed beyond recognition over the past 150 years. However, even if Marx can offer important insights for understanding modern capitalism, what about his claim that communism is the future of humanity? Surely the collapse of the Soviet bloc, China's economic reforms, and the implosion of left organisations across the world prove that Marx was wrong?

Contributors to this book beg to differ. *Anti-Capitalism: A Marxist Introduction* explains the structural features and the main shortcomings of modern capitalism, in order to substantiate our case against capitalism as a *system*. Chapters 1, 2 and 3 show that Marx's value theory provides important insights for understanding the modern world, including the exploitation of the workers, the sources of corporate power and the sickening extremes of overconsumption and widespread poverty. Chapters 5, 10 and 17 claim that classes exist, and that class struggle is, literally, alive and kicking around us. Chapters 4 and 6 show that technical change is not primarily driven by the urge to produce cheaper, better or more useful goods, but by the imperatives of profit-making and social control. Chapter 8 reviews the driving forces of capitalism across history, and Chapter 7 shows that capitalism is inimical to the Earth's ecological balance. Whereas environmental sustainability demands a very long-term calculation of costs and benefits, capitalism is based on short-term rationality and profit maximisation. *This social system must be confronted, in order to preserve the possibility of human life on this planet.*

Chapters 9 to 16 challenge other idols of contemporary thought, including the claims that capitalism promotes democracy, world peace and equality within and between nations, that every debt must be paid, that globalisation is unavoidable and unambiguously good, that national states are powerless, and that economic crises can be eliminated. Finally, Chapters 18 and 19 argue that capitalism is both unsustainable and undesirable. In our view, communism is justified not only on material but, especially, on human grounds. Much of what we argue is obvious. Yet often the obvious must be demonstrated over and over again, until it becomes self-evident to the majority.

This book also challenges the knee-jerk reaction against critiques of contemporary capitalism, the trite motto that 'there is no alternative' (TINA). Leading proponents of TINA include rapacious free-marketeers, prematurely aged philosophers of the 'Third Way', delusional economists, opportunistic politicians, corrupt bureaucrats, bankrupt journalists and other desperados. They claim that human beings are genetically programmed to be greedy, that capitalism is the law of nature, that transnational capital is usually right, and that non-intrusive regulation is possible when it goes wrong. They argue that capitalist societies, even though historically recent, will last forever, and that the triumph of the market should be embraced because it is both unavoidable and advantageous to all. They reassure us that massive improvements in living standards are just around the corner, and that only a little bit more belt-tightening will suffice.

Deceptions such as these have helped to legitimise the growing marketisation of most spheres of life in the last 20 years. In rich countries, this has taken place primarily through the assault on the social safety nets built after the Second World War. Low paid and insecure jobs have been imposed on millions of workers, the provision of public services has been curtailed, and the distribution of income and wealth has shifted against the poor. In poor countries, national development strategies have collapsed nearly everywhere. Under Washington's guidance, a bleak 'era of adjustment' has taken hold across the so-called developing world. In these countries, low expectations and policy conformity are enforced by usurious foreign debts and neoliberal policy despotism monitored by the IMF, the World Bank and the US Treasury Department. Recent experience abundantly shows that neoliberalism tramples upon the achievements, lives and hopes of the poor everywhere, and that it often leads to disastrous outcomes (see below).[3]

In spite of the much repeated claim that history is dead or, more precisely, that significant social and political changes are no longer possible, the neoliberal-globalist project has been facing difficult challenges. It has suffered legitimacy problems in the United States because of falling wages in spite of rising national income, in Western Europe because of simmering social conflicts triggered by high unemployment and stagnant living standards, and in Japan because of the protracted economic crisis. It has had to contend with the social and economic collapse of the former Soviet bloc, and with repeated financial and balance-of-payments crises in South East Asia and Latin America. It has also had to explain away the economic and political meltdown in sub-Saharan Africa, and to face frequent wars and unprecedented levels of terrorist activity across the world. Last but not least, neoliberal globalism has been confronted by profound disillusion everywhere, and by vibrant protests and mass resistance, especially in Argentina, Ecuador, Indonesia, Mexico, the Occupied Territories and South Korea.

In this context, the recent 'anti-globalisation' or 'anti-capitalist' protest movements are important for two reasons. First, they are global in scope, combining campaigns that were previously waged separately. In doing so, they have raised questions about the *systemic* features of capitalism for the first time in a generation. Second, they have shed a powerful light upon the dismal track record of contemporary capitalism. Although initially marginalised, these movements shot to prominence in the wake of the Zapatista rebellion, the Jubilee 2000 campaign and the confrontations that brought to a halt the Seattle WTO meeting. The new movements have joined vigorous mass demonstrations in several continents, and they have shown their opposition to the monopolistic practices of the TNCs, including pharmaceutical giants and corporations attempting to force-feed the world with genetically modified crops. They have challenged patent laws and clashed against other forms of 'corporate greed', leading to boycotts against Shell, Nike and other companies. These movements have also targeted repressive regimes, such as Myanmar's military dictatorship, and shown international solidarity, for example, with the Zapatistas and the Brazilian landless peasants.

In spite of their rapid growth, these movements remain fragmented. Different organisations pursue widely distinct objectives in diverse ways, and occasionally come into conflict with one another. The lack of a common agenda can hamper their ability to challenge established institutions and practices. Several pressure groups, including the environmental, peace, women's, gay, lesbian, anti-racist and animal liberation movements, international solidarity organisations, trade unions, leftist parties and other groups, defend their autonomy vigorously, sometimes allowing sectional interests to cloud their mutual complementarity. In spite of these limitations, political maturity, organisational flexibility and heavy use of the internet have allowed the new movements to expand. Moreover, they have often been able to transcend the rules, habits and conventions that constrain the NGOs, trade unions, political parties and other institutions of the left. Their recent successes show that there is widespread discontent and fertile ground for the discussion of alternatives, at different levels, around the world.

Continuing confrontation against the neoliberal-globalist project and its destructive implications is inevitable. Perhaps more significantly, it is likely that the anti-capitalist feeling previously channelled through trade unions and political parties of the left has found new outlets. If true, this shift will have important implications for the political landscape.

11 SEPTEMBER AND BEYOND

The growing opposition to the neoliberal–globalist project was temporarily checked by the tragic events of 11 September 2001. In response to those terrorist atrocities, the US government has unleashed a loosely targeted state terrorist campaign against millions of people, both at home and abroad. The most important thrust of this strategy has been the so-called 'infinite war' against elusive (but always carefully selected) adversaries. Rather than helping to resolve existing grievances, US state terrorism has provided further excuses for private terrorists around the world to target the United States and its citizens. In our view, all forms of terrorism – whether private, state-sponsored or state-led – are reactionary, repulsive, destructive, criminal and utterly unacceptable.

The so-called 'war on terror' has been rationalised by the naked conflation between the neoliberal-globalist agenda and US imperialism. The global elite (the Washington-based 'international community') has brazenly subordinated international law to US foreign policy interests. It has granted itself a licence to apply unlimited force against unfriendly regimes ('rogue states') or social movements ('terrorist organisations'), either for so-called humanitarian reasons or in order to defeat whatever it decides to call 'terrorism'.[4]

The overwhelming military superiority of the United States allows its government to pound foreign adversaries anywhere, secure in the knowledge that its own casualties will be small and that the damage to the other side will eventually crack the opposition. The war unleashed by the United States and its vassal states against Iraq, in 1990–91, and further military action in Afghanistan, Bosnia, Kosovo, Palestine, Panama, Sierra Leone, Somalia, Sudan and elsewhere have brought important gains to the global elite, not least unprecedented security guarantees for its business interests. However, the cost of these operations is incalculable. Conveniently, the victims are almost invariably dark-skinned and poor. They speak incomprehensible languages and worship lesser gods. They live in intractable trouble spots, which they are rarely allowed to leave because (in contrast with their money and goods) they are not welcome abroad. Their fate is of little concern, as long as they ultimately comply with Western geopolitical designs.

The tragedy of 11 September has exposed the limits of neoliberal globalism. The depth of dissatisfaction with Washington's political and economic rule has been revealed, and the claim that trade and financial liberalisation can resolve the world's most pressing problems has suffered a severe blow. The argument that states are powerless against the forces of globalisation has been undermined by the expansionary economic policies adopted in the wake of the attacks, and by the co-ordinated wave of repression unleashed across the world. Repression included not only the restriction of civil liberties, but also refined controls against capital flows and the limitation of property rights, for example, against pharmaceutical patents in the United States at the height of the anthrax threat. Finally, important anti-war movements emerged in several countries, especially the United Kingdom, Italy and – courageously – the United States.

In the wake of the tragedy of 11 September, the global elite seized the opportunity to open its batteries against all forms of dissent. Amid a rising tide of xenophobia and racism, rabid journalists cried out that anti-corporate protests were also anti-American, and scorned principled objections against the 'war on terror'. Colourful politicians on both sides of the Atlantic, eager to please their masters, even claimed that the new protest movements share the same objectives as Osama bin Laden.[5]

Difficulties such as these bring to the fore the need for clarity of objectives and careful selection of targets when campaigning against important features or consequences of modern capitalism. Unless our objectives are clear and the instruments appropriate, we will be unable to achieve our goals, at great cost to ourselves and the world.

Four issues play critical roles in the analysis of contemporary capitalism and, consequently, in the search for alternatives: neoliberalism, globalisation, corporate power and democracy. It is to these that we now turn.

FOUR PRESSING ISSUES

Neoliberalism

In the last 20 years, for the first time in history, there has been a concerted attempt to implement a single worldwide economic policy, under the guise of neoliberalism. The IMF, the World Bank, the US Treasury Department and, more recently, the European Central Bank (ECB), have strongly campaigned for neoliberalism, and they have sternly advised countries everywhere to abide by their commands. In this endeavour, they have been supported by the mainstream media, prestigious intellectuals, bankers, industrialists, landowners, speculators and opportunists vying for profits in every corner of the globe.

The spread of neoliberalism is due to several factors. These include the rise of conservative political forces in the United States, the United Kingdom and other countries, and the growing influence of mainstream theory within economics, both in its traditional form and through new institutionalism.[6] The forward march of neoliberalism was facilitated by the perceived failure of Keynesianism in the rich countries and developmentalism in poor ones, and by the collapse of the Soviet bloc. Finally, the US government has leaned heavily on the IMF, the World Bank, the United Nations and the World Trade Organisation (WTO) to promote neoliberal policies everywhere. Pressure by these organisations has validated the increasing use of aid, debt relief and foreign investment as tools with which to extract policy reforms from foreign governments.

Neoliberal policies are based on three premises. First, the dichotomy between markets and the state. Neoliberalism presumes that the state and the market are distinct and mutually exclusive institutions, and that one expands only at the expense of the other. Second, it claims that markets are efficient, whereas states are wasteful and economically inefficient. Third, it argues that state intervention creates systemic economic problems, especially resource misallocation, rent-seeking behaviour and technological backwardness.

These premises imply that certain economic policies are 'naturally' desirable. These include, first, rolling back the state in order to institute 'free markets', for example, through privatisation and deregulation of economic activity. Second, tight fiscal and monetary policies, including tax reforms and expenditure cuts, in order to control inflation and limit the scope for state intervention. Third, import liberalisation and devaluation of the exchange rate, to promote specialisation according to comparative advantage, stimulate exports and increase competition in the domestic market. Fourth, liberalisation of capital flows, to attract foreign capital and increase domestic capacity to consume and invest. Fifth, liberalisation of the domestic financial system, to increase savings and the rate of return on investment. Sixth, labour market flexibility, to increase the level of employment. Seventh, overhauling the legal system, in order to create or protect property rights. Eighth, political democracy, not in order to safeguard freedom and human rights but, primarily, to dilute state power and reduce the ability of the majority to influence economic policy.

It has been obvious for many years that these policies are successful only exceptionally. Economic performance during the last 20 years, in rich and poor countries alike, has been disappointing, with growth rates usually lagging behind those in the preceding (Keynesian) period. Poverty levels have not declined significantly, if at all; inequality within and between countries has increased substantially; large capital flows have been associated with currency crises, and the fêted economic transition in the former Soviet bloc has been an abysmal failure (at least for the majority). Neoliberals invariably claim that these disasters show the need for further reform. However, it is equally logical, and more reasonable, to conclude that the neoliberal reforms share much of the blame for the dismal economic performance in rich as well as poor countries.

The above conclusion is reinforced by five theoretical arguments.[7] First, neoliberal reforms introduce policies that destroy large numbers of jobs and entire industries, tautologically deemed to be 'inefficient', whilst relying on the battered patient to generate healthy alternatives through the presumed efficacy of market forces. This strategy rarely works. The depressive impact of the elimination of traditional industries is generally not compensated by the rapid development of new ones, leading to structural unemployment, growing poverty and marginalisation, and to a tighter balance-of-payments constraint in the afflicted countries.

Second, neoliberal faith in the market contradicts even elementary principles of neoclassical economic theory. For example, in their 'second best analysis', developed half a century ago, Lipsey and Lancaster demonstrate that, if an economy departs from the perfectly competitive ideal on several counts (as all economies invariably do), the removal of one 'imperfection' may not make it more efficient. Therefore, even mainstream economic theory can explain why neoliberal reforms can be worse than useless.

Third, the presumption that the market is virtuous while the state is wasteful, corrupt and inefficient is simply wrong. This false dichotomy is often employed in order to justify state intervention on behalf of capital (for example, privatisation and the curtailment of trade union rights facilitate capitalist abuse, consumer 'fleecing' and the increased exploitation of the workforce). In fact, states and markets are both imperfect and inseparable. They include many different types of institutions, whose borders cannot always be clearly distinguished. For example, the inland revenue service, financial services regulatory agencies, accounting and consultancy firms and state-owned and private banks are inextricably linked to one another, but the precise nature of their relationship is necessarily circumstantial.

Fourth, economic policies normally do not involve unambiguous choices between state and markets but, rather, choices between different forms of interaction between institutions in the two spheres. Privatisation, for example, may not imply a retreat of the state or even increased efficiency. The outcome depends on the firm, its output, management and strategy, the form of privatisation, the regulatory framework, the strength and form of competition, and other factors.

Fifth, developed markets arise *only* through state intervention. The state establishes the institutional and regulatory framework for market transactions, including property rights and law enforcement. It regulates the provision of infrastructure, ensures that a healthy, trained and pliant workforce is available, and controls social conflict. The state establishes and regulates professional qualifications and the accounting conventions, and develops a system of tax collection, transfers and expenditures that influences the development of markets, firm performance, and employment patterns. Since capitalist economies rely heavily and necessarily on state institutions, attempts to measure the degree of state intervention are simply misguided. What *really* matters is the gains and losses for each type of state policy, and the implementation of purposeful and co-ordinated policies.

This approach to markets and states does not deny the Marxian claim that the state is 'a committee for managing the common affairs of the whole bourgeoisie'[8] or that it is 'an essentially capitalist machine ... the state of the capitalists, the ideal collective body of all capitalists'.[9] The reasons are easy to understand. First, the state is *constitutionally* committed to capitalism by custom and law, and state institutions are geared towards, and have been historically shaped by, the development of markets, wage employment and profit-making activities. Second, the staffing and policy priorities of the state institutions are heavily influenced by the interest groups represented in and through them, where capital tends to be hegemonic. Third, the reproduction of the state relies heavily on the fortunes of capital, because state revenue depends upon the profitability of enterprise and the level of employment. Fourth, the economic and political power of the capitalists, and their influence upon culture, language and habits, is overwhelming, especially in democratic societies. For example, although the commodification of votes, state control of the media and the imposition of openly ideological selection criteria for state officials are usually associated with the strong-arm tactics of African chiefs and Latin American landlords, they are nowhere more prominent than in the United States.

In conclusion, economic policy and its effects are both context-dependent *and* structured by the needs of capital. On the one hand, pressure for or against specific policies *can be effective*, and the ensuing policy choices *can* improve significantly the living conditions of the majority. On the other hand, these potential successes are limited. When faced with 'unacceptable' policies, the capitalists will refuse to invest, employ, produce and pay taxes; they will trigger balance-of-payments crises, cripple the government, paralyse the state and hold the workers to ransom. And they will not hesitate to resort to violence to defend their power and privileges. History abundantly shows that most state institutions, including the police and the armed forces, will rally around the moneyed interests and seek to protect them against challenges from below.

Globalisation

'Hyper-globalism' is the international face of neoliberalism. During the 1990s, analysts and pundits stridently claimed that developments in technology, communications, culture, ideology, finance, production, migration and the environment have modified the world beyond recognition. Drawing on these superficial insights, the 'hyper-globalists' argue that globalisation entails the supremacy of international over domestic institutions, the decline of state power, and the relentless domination of social life by global markets.[10]

Neoliberals have been at the forefront of the hyper-globalist assault. Most neoliberals proclaim both the virtues and the inevitability of the coming world market for everything (except labour, to be kept caged behind borders). They argue that markets ought to reign unimpeded by national legislation and meddling international organisations and, implausibly, claim that policy subordination to *global* imperatives is essential for *national* welfare.

Hyper-globalist views have been discredited by a range of critical studies. These studies show, first, that global integration builds upon, rather than denies, the existence of nation states, which remain the seat of legitimacy and political and economic power. Rather than withering away because of the penetration of TNCs, vast international capital flows and the weight of international treaties, the critics have argued that powerful states promote international integration in pursuit of their own agendas, especially improved competitive positions for home capital in key business areas. Second, global neoliberalism has been associated with undesirable outcomes, including increasing poverty and inequality, the debasement of democracy and the erosion of the welfare state, to the benefit of powerful corporations and financial interests. Third, the critical literature claims that globalisation is neither new nor overwhelming. It was preceded by similar episodes, especially before the First World War; it is not truly 'global', being largely restricted to trade and investment flows between developed countries and, even in this restricted sphere, capital is not 'free' to move at will; finally, in spite of appearances to the contrary, the net macroeconomic effect of trade and financial liberalisation is often very small. Fourth, the critics argue that the hyper-globalists conflate 'global' markets with the theoretical construct of perfect competition, characterised by perfect information and costless capital mobility. This confusion provides ideological cover for pro-business policies and for aggressive state intervention to foster private capital accumulation.

These critiques of hyper-globalism have led to three policy conclusions, which may or may not be mutually compatible. Some have argued for 'localisation', or the decentralisation of the world economy with increasing reliance on local production and exchange. Others have emphasised the need to democratise policy-making, including an increased role for sector-specific trade and industrial policy and national controls on capital flows. Yet others have pursued 'internationalisation', or the reform and revitalisation of international institutions (the UN, IMF, World Bank, WTO, EU, ECB, and so on), in order to promote the positive aspects of globalisation.[11]

Unfortunately, there are severe problems with each of these alternatives. 'Localisation' promotes small capital vis-à-vis large capital, represented by TNCs. This can be analytically misguided, because it ignores the close relationship that exists between large and small firms. For example, small firms often cluster around and supply parts and other inputs to large firms, provide cleaning and maintenance services, and so on. Their relationship can be so close as to render 'separation' between these firms impossible. Moreover, small firms tend to be financially fragile, lack the resources for technical innovation and the adoption of new technologies developed elsewhere, cannot supply large markets, and often treat their workforces more harshly than large firms. Finally, curbing the TNCs will inevitably reduce the availability of important commodities across the globe, including foodstuffs, electronic appliances and industrial machinery.

Attempts to 'recover' industrial policy for progressive ends can be successful; however, misguided policies can be useless and even counterproductive. Finally, 'internationalisation' is utopian. Most international institutions are firmly under the grip of the neoliberal-globalist elites, and it is unrealistic to expect that control can be wrested from them. In most cases, these institutions ought to be abolished, to be replaced, when necessary, by alternatives designed from scratch.

The insufficiencies of these critiques of hyper-globalism are often due to the misguided opposition between the global, national and local spheres. This separation mirrors that between markets and states, discussed above. In general, those spheres should not be contrasted as if they were mutually exclusive, because they constitute one another and can be understood only through their mutual relationship.

Specifically, the presumption that the local and national economies are the building blocs of the global economy is misguided. The so-called 'global' economy is nothing but the commuters daily going to the Manhattan financial district and the City of London, manual workers clocking into position in the Ruhr, English-speaking call-centre workers cycling to their jobs in Mumbai, stevedores working in Maputo, and hundreds of millions of workers producing for people living in distant lands, and consuming not only locally produced goods but also commodities produced elsewhere. In this sense, there is little difference between domestic and cross-border economic transactions, and economic growth necessarily encompasses the simultaneous development of the local, national *and* global economies. In fact, there are reasons to believe, first, that important aspects of production and finance have always been 'international'. Second, that long-distance trade has been *more* important for social and economic development than exchanges between neighbours. Third, that capitalism originally developed neither in a single country nor in discrete regions, but locally, regionally and internationally *at the same time.*

Terms like 'globalisation' or the 'internationalisation of production and finance', on their own, are simply *devoid of meaning.* Capital is neither national nor international; it is a relationship between people that appears as things or money. Consequently, there is nothing intrinsically national *or* international about capitalist institutions, production or practices. Detailed studies have shown, for example, that 'globalisation' is not a homogeneous, unidirectional and inevitable process taking place between neatly separated national economies. Globalisation does not tend to 'eliminate' the nation state, and recent developments in production, finance, culture, the environment, and so on are profoundly different from one another and must be analysed separately. What is often called 'globalisation' is, in fact, a set of more or less interlocking processes, some articulated systemically and others largely contingent, moving at different speeds and in different directions across different areas of the world economy. Some of these processes tend to erode national states and local identities, while others reinforce them.

Both wholesale support for 'globalisation' and wholesale opposition to it are profoundly misguided (for example, it makes no sense for a *global* protest movement to be called 'anti-globalisation'). What matters, at the local, national *and* global levels, is what is produced and how, by whom, and for whose benefit. In the early twenty-first century, as in the mid-nineteenth century, the distances between people matter less than the relationships between them. Similarly, geography remains less important than the social structures of control and exploitation that bind people together within cities, between regions, and across the world.

Corporate power

The new 'anti-capitalist' movements are famously critical of the large corporations, especially TNCs. This section argues that the market power and political influence of TNCs raise important ethical and economic questions. However, TNCs are not new, and their recent expansion is not the harbinger of fundamental changes in the economic and political landscape. Therefore it would be misguided to try to turn them into the main focus of resistance.

Several commentators sympathetic to the new movements claim that one of the most important problems of contemporary capitalism is the excessive tilting of power towards the large corporations. The causes and implications of this process are usually left unexamined, although they are presumably related to neoliberalism and globalisation. It is also left unclear what should be done about it, other than imposing unspecified curbs against corporate power.

This is clearly insufficient. Arguments along those lines are often fruitless because they are not based on a consistent theory of the state and its relationship to the corporations, and on a theory of monopoly power and capitalist behaviour, without which corporate practices cannot be understood. For example, although it is right to claim that the state is controlled by capitalist interests and forces (see above), it is wrong to ascribe boundless power to specific groups or interests, such as the TNCs, financiers, landlords or foreign capitalists. No social group can exist in isolation, and none exercises unlimited power.

Let us analyse in more detail the claim that 'large firms' control production, exchange, distribution and the political process. This view is incorrect for four reasons. First, it artificially disassembles capital into 'large' and 'small' units (see above). Second, it suggests that small firms, such as tiny grocery stores, family-owned newsagents and small farms conform more closely to local interests, as if they were independent of the large firms which they represent and that provide them with inputs and markets, and as if small firms were renowned for their promotion of employee interests. Third, it erroneously implies that the evils of capitalism are due to the large firms only, and that these wrongs can be put right by anti-monopoly legislation and domestic market protection against foreign firms. Fourth, this view misrepresents 'competitive capitalism', as if it had actually existed at some idyllic point in the past. In this idealised image of Victorian capitalism, unsightly features such as poverty, imperialism, slavery, genocide and the forces that transformed 'competitive' into 'monopoly' capitalism are arbitrarily expunged.

Sleights of hand such as these, and the lack of a theory of capital, the state, competition and monopoly power, explain the coexistence of critiques of corporate practices with pathetic apologias for capitalism. For example, in the words of a well-known critic of 'globalisation':

> My argument is not intended to be anti-capitalist. Capitalism is clearly the best system for generating wealth, and free trade and open capital markets have brought unprecedented economic growth for most if not all of the world. Nor is ... [it] anti-business ... [U]nder certain market conditions, business is more able and willing than government to take on many of the world's problems ... I mean to question the moral justification for a brand of capitalism ... in which we cannot trust governments to look after our interests in which unelected powers – big corporations – are taking over governments' roles.[12]

This approach is profoundly misguided. The outrageous behaviour of large corporations, from the East India Company to Microsoft, and from ITT to Monsanto, is not primarily due to their size, greed, or the support of states that they have hijacked at some mysterious point in time. Corporate practices and monopoly power are due to the forces of *competition*. By the same token, our collective addiction to McChickens and corporate logos is not simply due to the crude manipulation of our desires by brutish TNCs. Corporate behaviour (and its welfare implications), is ultimately rooted in the dominance of a system of production geared towards *private profit* rather than *collective need*.

Democracy

Several critics have recently highlighted the increasing emasculation of democracy, the erosion of citizenship and the declining account-ability of the state even in 'advanced' democratic societies. These processes are often blamed on the capture of the state by corporate and other interest groups. However, this view is misleading, and the explanation is inadequate.

This section briefly reviews the relationship between the state, capital, the political regime and economic policy. Along with most of the literature, it claims that political freedom is immensely valuable, and that the spread of democracy across the world has been possible only through the diffusion of capitalism. However, this section also shows that *capitalism necessarily limits democracy*, and that the expansion of democracy into critically important areas of life requires the *abolition* of capitalism.[13]

A remarkable distinction between precapitalist and capitalist societies is the separation, in the latter, between the 'economic' and 'political' spheres. This separation means that, under capitalism, 'economic' processes – including the production, exchange and distribution of goods and services, the compulsion to work and the exploitation of the workers – are generally carried out 'impersonally', through market mechanisms. It is completely different in precapitalist societies. In these societies, economic processes are directly subordinated to political authority, including both personal command and state power, and they generally follow rules based on hierarchy, tradition and religious duty.

The separation between the economic and political spheres has three important implications. First, it leads to the constitution of a separate 'political' sphere. For the first time in history, the owners of the means of production are relieved from public duty, which becomes the preserve of state officials. The separation of the political sphere establishes the potential and limits of state intervention in the economy, including the scope of economic policy and the possibility of 'autonomous' political change, with no direct implication for the 'economic' order. The substance and degree of democracy is a case in point (see below).

Second, separation entrenches capitalist power within the 'economic' sphere. Manifestations of economic power include the ownership and control of means of production (the factories, buildings, land, machines, tools and other equipment and materials necessary for the production of goods and services), the right to control the production process and discipline the workforce, and the ability to exploit the workers.

Third, the separation between the economic and political spheres is relative rather than absolute. On the one hand, the 'political' power of the state and the 'economic' power of the capitalists may lead to conflict, for example, over the conditions of work, the minimum wage, pension provisions and environmental regulations. On the other hand, we have already seen that modern states are essentially *capitalist*. Experience shows that the state will intervene directly both in 'political' conflicts (for example, the scope of democratic rights) and in purely 'economic' disputes (for example, pay and conditions in large industries), if state officials believe that their own rule or the reproduction of capital are being unduly challenged. When intervening, the state relies on the power of the law, the police and, *in extremis*, the armed forces.

The existence of a separate political sphere, explained above, implies that capitalism is compatible with political (formal or procedural) democracy. Political democracy includes the rule of law, party-political pluralism, free and regular elections, freedom of the press, respect for human rights, and other institutions and practices that are essential for the consolidation of human freedom.

However, capitalism *necessarily* limits the scope for freedom because it is inimical to *economic* (substantive) democracy. These limits are imposed by the capitalist monopoly over the economic sphere, explained above. For example, the franchise and political debate are not generally allowed to 'interfere' with the ownership and management of the production units and, often, even with the composition of output and the patterns and conditions of employment, in spite of their enormous importance for social welfare. In other words, even though political campaigns can achieve important transformations in the property rights and work practices, the scope for democratic intervention in the economic sphere is always limited.

The limits of capitalist democracy come into view, for example, when attempts to expand political control over the social affairs are constrained by the lack of economic democracy – typically, when governments or mass movements attempt to modify property rights by constitutional means. The resulting clashes were among the main causes of the defeat of the Spanish Republic, the overthrow of Chilean president Salvador Allende and, less conspicuously but equally significantly, the systematic failure of attempted land reforms across Latin America. Mass movements attempting to shift property rights by legal means but against the interests of the state have also been crushed repeatedly, in many countries. In these clashes, the success of the conservative forces often depends upon the arbitrary limitation of political democracy. This implies that political democracy is rarely able to challenge successfully the economic power of the capitalist class (embodied in their 'core' property rights). This is not a matter of choice: *the advance of political democracy is permanently limited by the lack of economic democracy.*

Tensions between economic and political democracy generally surface through the ebb and flow of political democracy and civil rights. These tensions are nowhere more visible than in the 'developing' countries. In recent years, multi-party democracy and universal suffrage have been extended across the world, the repressive powers of the state have been curtailed by the United Nations and the International Court of Justice, and by the precedents established by the Pinochet affair and the prosecution of officials of the former Rwandan government.

In spite of these important advances, the forward march of political democracy has been severely hampered by the exclusion of economic matters from legitimate debate. The imposition of neoliberalism across the world is the most important cause of these limitations. Because of neoliberalism, worldwide policy-making capacity has been increasingly concentrated in Washington and in Wall Street, leaving only matters of relatively minor importance open for debate, both in 'developing' and developed countries.

Specifically, in the 'newly democratic' states of Latin America, sub-Saharan Africa and South East Asia the transitions towards political democracy were generally conditional upon compromises that ruled out substantive shifts in social and economic power. Even more perversely, in these countries *the imposition of neoliberal policies often depended upon the democratic transition*. After several decades attempting to subvert democratic governments and shore up dictatorships across the globe, the US government and most local elites have realised that *democratic* states can follow diktats from Washington and impose policies inimical to economic democracy more easily and reliably than most dictatorships. This is due to the greater *political legitimacy* of formally democratic governments.

This argument can be put in another way. Repression is often necessary in order to extract the resources required to service the foreign debt, shift development towards narrow comparative advantage and support parasitical industrial and financial systems. However, dictatorships can rarely impose the level of repression necessary to implement neoliberal policies. This is something that only democratic states can do successfully, because *their greater legitimacy allows them to ignore popular pressure for longer* (however, the recent upheavals in Argentina show that this strategy is also limited).

In this sense, the neoliberal-globalist project involves a fundamental inconsistency: it requires *inclusive* political systems to enforce *excluding* economic policies. These policies demand states hostile to the majority, even though democratic states are supposedly responsive to democratic pressure. As a result, we see across the world the diffusion of *formally democratic* but *highly repressive* states. We also see the perpetuation of social exclusion and injustice, in spite of political pluralism and the consolidation of democratic institutions in many countries.

'Democratic neoliberalism' has consolidated *economic apartheid* both within and between countries. Economic apartheid includes the increasing concentration of income and wealth, the segregation of the upper classes in residential, work and leisure enclosures, their unwillingness and inability to interact with the poor in most spheres of social and civic life, the diffusion of organised and heavily armed criminal gangs, and unbridled corruption in state institutions.

Economic apartheid and the evacuation of economic democracy can be at least partly reversed through successful mass struggles. These struggles can limit the power of industrial and financial interests, and open the possibility of policy alternatives leading to improvements in the living conditions of the majority. However, democracy can be extended into critically important spheres of life *only* if the capitalist monopoly over the economic sphere is abolished. In this sense, the success of the struggle depends on the extent to which the democratic movement becomes *anti-capitalist.*

THE WAY AHEAD

The previous section has shown that we should not expect significant transformations of contemporary capitalism through appeals for the restoration of state power, the reform of international institutions, campaigns for corporate responsibility or the expansion of formal democracy. Reforms are certainly possible in these and in other areas, and they can increase greatly the power and influence of the majority. However, these reforms are always limited and, even if successful, they will be permanently at risk because they fail to address the root cause of the problems of contemporary capitalism.

Strategic success depends on five conditions. First, *holism.* Successful challenges against different forms of discrimination, 'shallow' democracy, the inequities of debt, the destructive effects of trade and capital flows, environmental degradation, corporate irresponsibility, and so on, require the consolidation of sectoral struggles into a single mass movement against the global rule of capital – the root cause of these wrongs.

Second, whilst the movement ought to remain *international*, it should focus its energies in the *national* terrain. This is only partly because the potential efficacy of the struggle is maximised at this level (it is much harder to mobilise successfully in the international sphere). It is also because national states play an essential role in the choice and implementation of economic policy, the operation of markets and the limitation of corporate power. Moreover, 'global capitalism' is organised primarily nationally, and its actors (TNCs, international organisations, global markets, and so on) depend heavily upon state promotion and regulation.

It was shown above that *there is no such thing as global capitalism* independently of national states and local workers and capitalists. By the same token, the most effective means of influencing 'global' developments is by exercising pressure upon national states. In fact, it is because the national states are the critical and, at the same time, the weakest links in the 'global economy' that capital endlessly repeats the myth that globalisation renders the state powerless and irrelevant.[14]

Third, the movement should develop further the *ability to mobilise large numbers of people by non-traditional means*, and pursue innovative forms of struggle.

Fourth, the growth of the movement depends heavily upon its ability to *incorporate the immediate concerns* of the majority. These include issues related to unemployment and overwork, low pay, lack of employment security and rights in the workplace, the degradation of heavily populated environments, the provision of public health, sanitation and clean and efficient transport and energy, and so on. Success also requires closer attention to the *workplace*, which is the basis of capitalist domination and economic power. Unity between economic and political struggles, and challenges against both capital and the state, especially through mass confrontation against *state economic policy* and its consequences, are important conditions for growth and victory.[15]

Fifth, given the limits of political democracy and state power, the achievement of equality and the elimination of poverty and exploitation within and between countries demands *transcendence*, or the abolition of capitalism. These conclusions are explained and substantiated by every chapter in this book.

LEAVING CAPITALISM BEHIND

Social reformers, utopian socialists, anarchists, social democrats, Marxists and many others have questioned the legitimacy and desirability of capitalism for at least two centuries. However, it is beyond dispute that Marxism provides the basis for the most comprehensive critique of this social and economic system, including the development of the radical alternative to capitalism: communism. The

Marxist analysis of transcendence can be divided into two areas: the critique of capitalism and the importance of communism.

Several problems of contemporary capitalism have been discussed above and, in each case, the root cause of these problems and the limits to their potential solution under capitalism were highlighted. Some of these problems can be remedied within the current system, for example, the erosion of political democracy, lack of corporate responsibility, and absolute poverty. In contrast, other problems cannot be resolved, because they are *features* of capitalism; among them, unemployment, exploitation of the workforce, economic inequality, the encroachment of work upon free time, systematic environmental degradation, the lack of economic democracy, and production for profit rather than need. Problems such as these can, at best, be concealed by propaganda and mitigated by economic prosperity.

Marxists claim that the limitations of capitalism can be eliminated only through the institution of another form of social organisation, communism. The misrepresentation of communism in the past two centuries cannot be put right in this book. However, three comments are in order. First, communism should not be confused with the political systems associated with the USSR or China.[16] Second, communism is neither inexorable nor unavoidable. Capitalism will change and, ultimately, be displaced, only if overwhelming pressure is applied by the majority. Failing that, capitalism may persist indefinitely, in spite of its rising human and environmental costs. Third, communism is neither an earthly version of paradise, nor the 'end of history'. Quite the contrary: communism marks the end of the *prehistory* of human society. Communism will eliminate the socially created constraints of poverty, drudgery, exploitation, environmental degradation, and other limitations currently caused by the manic search for profit. Removal of these constraints will allow history to *begin*, because human beings will, finally, free themselves from the dictatorship of moneyed interests, destitution due to the existence of large-scale property, and inequality engendered by wealth and privileged upbringing. *Economic equality is essential for political equality*, thus allowing everyone to become a valued member of a truly open society.

The struggle against capitalism is part and parcel of the struggle for democracy in society and in the workplace, against profit and privilege, and for equality of opportunity for everyone. These are the struggles that define the new movements, but taken to their logical conclusion.

REFERENCES

Arestis, P. and Sawyer, M. (1998) 'New Labour, New Monetarism', *Soundings*, Summer; reprinted in *European Labour Forum* 20, Winter, 1998–99.

Barker, C. (2001) 'Socialists', in E. Bircham and J. Charlton (eds.) *Anti-Capitalism: a Guide to the Movement*. London: Bookmarks.

Callinicos, A. (2001) 'Where Now?', in E. Bircham and J. Charlton (eds.) *Anticipation: A Guide to the Movement*. London: Bookmarks.

Chattopadhyay, P. (1994) *The Marxian Concept of Capital and the Soviet Experience: Essay in the Critique of Political Economy*. Westport, Conn.: Praeger.

Engels, F. (1998) *Anti-Duhring*, CD-Rom. London: Electric Books.

Fine, B. (2001) *Globalisation and Development: The Imperative of Political Economy*, unpublished manuscript.

Fine, B., Lapavitsas, C. and Pincus, J. (eds.) (2001) *Development Policy in the Twenty-first Century: Beyond the Post-Washington Consensus*. London: Routledge.

Fine, B. and Stoneman, C. (1996) 'Introduction: State and Development', *Journal of Southern African Studies* 22 (1), pp. 5–26.

German, L. (2001) 'War', in E. Bircham and J. Charlton (eds.) *AntiCapitalism: A Guide to the Movement*. London: Bookmarks.

Hertz, N. (2001) *The Silent Takeover: Global Capitalism and the Death of Democracy*. London: William Heinemann.

Karliner, J. (2001) 'Where Do We Go From Here? Pondering the Future of Our Movement', *CorpWatch*, 11 October 2001 (www.corpwatch.org).

Marx, K. and Engels, F. (1998) *The Communist Manifesto*, Cd-Rom. London: The Electric Book Company.

Radice, H. (2000) 'Responses to Globalisation: a Critique of Progressive Nationalism', *New Political Economy* 5 (1), pp. 5–19.

WDM (2000) *States of Unrest: Resistance to IMF Policies in Poor Countries*. London: World Development Movement (www.wdm.org).

Wood, E.M. (1981) 'The Separation of the Economic and the Political in Capitalism', *New Left Review* 127, pp. 66–95.

Wood, E.M. (1988) 'Capitalism and Human Emancipation', *New Left Review* 167, January–February, pp. 3–20.

Wood, E.M. (2002) 'Global Capital, National States', in M. Rupert and H. Smith (eds.) *Now More Than Ever: Historical Materialism and Globalisation*. London: Routledge, forthcoming.

NOTES

1. I am grateful to Ben Fine and Mike Lebowitz for their helpful comments and suggestions.
2. Marx and Engels (1998, pp. 13–14), emphasis added.
3. Resistance against IMF policies in poor countries is documented in WDM (2000).
4. See German (2001, pp. 126–7).

5. 'G7 activists no better than Bin Laden' (*London Evening Standard*, November 5, 2001). Similar claims were reportedly made by US Representative Don Young, US Trade Representative Robert Zoellick and Italian Prime Minister Berlusconi, among others (Karliner 2001).

6. See Fine, Lapavitsas and Pincus (2001).

7. See Arestis and Sawyer (1998) and Fine and Stoneman (1996), on which this section draws, and the references therein.

8. Marx and Engels (1998, p. 12).

9. Engels (1998, p. 352).

10. This section draws on the critical surveys by Radice (2000) and, especially, Fine (2001).

11. For a similar analysis, see Callinicos (2001).

12. Hertz (2001, p. 10).

13. For a detailed analysis, see Wood (1981).

14. See Wood (2002).

15. Barker (2001, p. 333) rightly argues that 'Putting a brick through the window of Starbucks is a moral gesture, but an ineffective one. Organising Starbucks workers is harder, but more effective – and hurts the Starbucks bosses more ... We need to focus on people's lives as producers and not simply as consumers – for there is a power in producers' hands that consumer boycotts can never match. In any case, many consumers can't afford to "choose".' Isaac Deutscher made a similar point to student activists in the mid-1960s: 'You are effervescently active on the margin of social life, and the workers are passive right at the core of it. That is the tragedy of our society. If you do not deal with this contrast, you will be defeated' (cited in Wood 1988, p. 4).

16. See Chattopadhyay (1994).

Part I

Capital, Exploitation and Conflict

1 Value, Capital and Exploitation[1]

Alfredo Saad-Filho

This chapter explains the essential elements of Marx's theory of value and exploitation.[2] This theory provides the foundation for his critique of capitalism, and it substantiates Marx's claim that capitalism is a historically limited system. Important elements of Marx's theory include his explanation of why wage workers are exploited, the sources of social conflict, the inevitability of, and systematic form taken by, technical change through the growing use of machinery, the determinants of wages, prices and distribution, the role of the financial system and the recurrence of economic crises.

COMMODITIES

If you lift your eyes from this page for a moment, you can see *commodities* everywhere. This book is a commodity and, in all likelihood, so are your other books, clothes and shoes, your TV, CD player, computer and other means of information and entertainment, and your home, bicycle, car and other means of transportation. Your beauty products are also commodities, and so are your holidays and food, including ready-made foods and the means to prepare food at home. Commodities are not only for individual consumption. At your place of work or study, most things are also commodities. You live in a world of commodities.

Commodities are goods and services produced for sale, rather than for consumption by their own producers. Commodities have two common features. On the one hand, they are *use values*: they have some characteristic that people find useful. The nature of their demand, whether it derives from physiological need, social convention, fancy or vice is irrelevant for our purposes. What matters is that commodities must be useful for others, making them potentially saleable.

On the other hand, commodities have *exchange value*: they can, in principle, be exchanged for other commodities (through money, see below) in specific ratios. For example, one small TV set is equivalent to one bicycle, three pairs of shoes, ten music CDs, one hundred cappuccinos, and so on. Exchange value shows that, in spite of their distinct use values, commodities are *equivalent* (at least in one respect) to one another. In this sense, in spite of their differences all commodities are the same.

In commodity economies (where most goods and services are commodities) money fulfils two roles. First, it simplifies the vast number of bilateral exchange ratios between these commodities. In practice, only the exchange value of commodities in terms of money (their price) is quoted, and this is sufficient to establish the equivalence ratios between all commodities. Second, commodity exchanges are usually indirect, taking place through money. For example, you do not produce all the goods and services that you want to consume. Rather, you specialise in the production of one commodity – say, restaurant meals, if you are a cook – and exchange it for those commodities that you want to consume. These exchanges are not direct (barter), as they would be if cooks offered their dishes to passers-by in exchange for cinema tickets, shoes, songs and automobiles. Instead, you sell your talents to a restaurateur in return for money and, armed with notes and coins (or a chequebook or bank card), you can purchase what you wish to consume (see Chapter 3).

LABOUR

The double nature of commodities, as use values with exchange value, has implications for labour. On the one hand, commodity-producing labour is *concrete labour*, producing specific use values such as clothes, food, books, and so on. On the other hand, as was shown above, when goods are produced for exchange (and have exchange value) they have a relationship of equivalence to one another. In this case, labour is also *abstract* (general) *labour*. Just like the commodities themselves, commodity-producing labour is both general and specific at the same time.

Concrete labour, producing use values, exists in every type of society, because people always and everywhere need to appropriate use values for their own reproduction – that is, to reproduce their own capacities as human beings. In contrast, abstract labour is historically specific; it exists only where commodities are being produced and exchanged.

Abstract labour has two distinct aspects – qualitative and quantitative – that should be analysed separately.

First, abstract labour derives from the relationship of equivalence between commodities. Even though it is historically contingent, abstract labour has *real* existence; it is not merely a construct of the mind. A visit to the local supermarket, for example, shows that your own labour is *actually* equivalent to the labours that have produced thousands of different goods, some of them nearby, and others halfway across the globe. Labours are equivalent (as abstract labour) because commodities are produced for exchange. Their equivalence appears through the convertibility between money and commodities. When you buy a chocolate bar, for example, you are realising the equivalence between your own labour – as a cook, for example – and the labour of the producers of chocolate. The ability of money to purchase any commodity shows that *money represents abstract labour.*

Second, the stability of the exchange values shows that there is a quantitative relationship between the abstract labours necessary to produce each type of commodity. However, this relationship is not direct, as we will see below.

In his *Inquiry into the Nature and Causes of the Wealth of Nations*, first published in 1776, Adam Smith claims that in 'early and rude' societies goods exchanged directly in proportion to the labour time necessary to produce them. For example, if 'it usually costs twice the labour to kill a beaver which it does to kill a deer, one beaver should naturally exchange for or be worth two deer' (Smith 1991, p. 41). However, Smith believes that this simple pricing rule breaks down when instruments and machines are used in production. The reason is that, in addition to the workers, the owners of the 'stock' also have a legitimate claim to the value of the product.

Marx disagrees with Smith, for two reasons. First, 'simple' or 'direct' exchange (in proportion to socially necessary labour) is not typical of any human society; this is simply a construct of Smith's mind. Second, and more importantly for our purposes, although commodity exchanges reveal the quantitative relations of equivalence between different types of labour, this relationship is indirect. In other words, whereas Smith abandons his own 'labour theory of value' at the first hurdle, Marx develops his own value analysis rigorously and systematically into a cogent explanation of commodity prices under capitalism (see below and, for details, Saad-Filho 2002).

CAPITALISM

Commodities have been produced for thousands of years. However, in non-capitalist societies commodity production is generally marginal, and most goods and services are produced for direct consumption by the household or for non-market exchange. It is different in capitalist societies. *The first defining feature of capitalism is the generalised production of commodities.* Under capitalism, most goods and services are produced for sale, most workers are employed in the production of commodities, and commodities are systematically traded in developed markets, where firms and households regularly purchase commodities as production inputs and final goods and services, respectively.

The second defining feature of capitalism is the production of commodities for profit. In capitalist society, commodity owners typically do not merely seek to make a living – they want to make *profit*. Therefore, the production decisions and the level and structure of employment, and the living standards of the society, are grounded in the profitability of enterprise.

The third defining feature of capitalism is wage labour. Like commodity production and money, wage labour first appeared thousands of years ago. However, before capitalism wage labour was always limited, and other forms of labour were predominant. For example, co-operation within small social groups, slavery in the great empires of antiquity, serfdom under feudalism, and independent production for subsistence or exchange, in all types of society. Wage labour has become the typical mode of labour only recently; three or four hundred years ago in England, and even more recently elsewhere. In some parts of the developing world, wage labour, complex markets and commodity production for profit still play only a minor role in social and economic reproduction.

WAGE LABOUR

Most people do not freely choose to become wage workers. Social and historical studies show that paid employment is generally sought only by those who cannot satisfy their needs in any other way. Historically, wage labour expands, and capitalist development takes off, only as the peasants, artisans and the self-employed lose control of the means of production (land, tools, machines and other resources), or as non-capitalist forms of production become unable to provide for subsistence (see Chapter 8).

The much-repeated claim that the wage contract is the outcome of a free bargain between equals is, therefore, both partial and misleading. Even though the workers are free to apply for one job rather than another, or to leave, they are almost always in a weak bargaining position when facing their prospective employers. Although they are not the property of individual employers, the wage workers need money in order to attend to the pressing needs of their household, including subsistence needs, mortgage and other debt payments and uncertainty about the future. These are some of the sticks with which capitalist society forces the workers to sign up 'freely' to the labour contract, 'spontaneously' turn up for work as and when required, and 'voluntarily' satisfy the expectations of their line managers (see Chapter 5).

The wage relation implies that the workers' capacity to work, their *labour power*, has become a commodity. The use value of the commodity labour power is its capacity to produce other use values (clothes, food, CD players, and so on). Its exchange value is represented by the wage rate. In this sense, labour power is a commodity like any other, and the wage workers are commodity sellers.

It is essential to distinguish between *labour* and *labour power*. Labour power is the *potential* to produce things, while labour is its *use* – in other words, labour is the *act* of transforming given natural and social conditions into a premeditated output (see Chapter 2). When a capitalist hires workers, she purchases the workers' *labour power* for a certain length of time. Once this transaction has been completed the workers' time belongs to the capitalist, who wishes to extract from them as much *labour* as possible within the terms of the contract. The workers, in turn, tend to resist abuse by the capitalist, and they may limit the intensity of labour unilaterally or reject arbitrary changes in the production norms. In sum, the purchase of labour power does not guarantee that a given quantity of labour is forthcoming, or that a certain quantity of value will be produced. The outcome depends upon persuasion and conflict in the shopfloor, farm or office.

MARKETS

The three features of capitalism (explained above) are not merely coincidental. There is a relationship of mutual determination between them. On the one hand, in advanced capitalist societies a large variety of commodities are produced for profit by millions of wage workers in thousands of firms. Many of these commodities are later purchased by those workers, who no longer can or wish to provide for themselves. Therefore, the spread of the wage relation fosters, simultaneously, the supply of as well as the demand for commodities.

On the other hand, the diffusion of wage labour and commodity exchanges stimulates the development of markets. For mainstream economic theory, markets are merely a locus of exchange, and they are essentially identical with one another: price changes affect both supply and demand, sexy adverts can help to sell anything, and the rest is up to the sales team. This is both partial and misleading. Markets are part of the institutions and channels of circulation that structure the systems of provision in the economy. Systems of provision are the chains of activity connecting production, exchange and consumption, ranging from the supply of basic inputs (crude oil, copper, cotton, cocoa, and so on) through to the manufacturing stage and, finally, the distribution of the finished commodities (aviation fuel, CD players, tee shirts, chocolate and other products). At certain stages in these chains, some commodities are marketed on a regular basis. The necessity of market exchange, and the form it takes, depend upon the features of each system of provision.[3]

Four conclusions follow. First, markets are not ideal structures of exchange, that can be judged to be more or less 'perfect' according to their degree of correspondence with a general model of perfect competition (as is presumed by mainstream economic theory). Although markets are essential for commodity production and the realisation of profits, they exist only concretely, and the markets for fuel, clothes, food, computers, labour power, money, credit, foreign currencies and other commodities can be profoundly different from one another.

Second, markets are structured not only 'internally', by the systems of provision, but also 'externally', by the social and economic constraints affecting production and exchange, such as law and the justice system, the transportation, storage and trading facilities, the international trade relations, the monetary, financial and tax systems, and so on.

Third, capitalist producers gauge demand only indirectly, through the purchasing power of their customers and the profitability of enterprise. This is why markets often fail to satisfy important needs (for example, effective prevention and treatment for the diseases of the poor, such as malaria) and, conversely, why luxury, wasteful or harmful goods and services are produced in large quantities (cosmetic surgery, advertising, cigarettes, and so on).

Fourth, markets are often the venue of vicious and wasteful struggles for profit. Reality does not correspond to mainstream theory, where market competition almost always is efficient and leads to optimum outcomes. In the real world, expensive advertising campaigns, employing large numbers of talented people, are regularly concocted to lure potential customers into purchasing whatever product the capitalists want to sell. Brand names are artificially differentiated, and virtually identical products compete wastefully for attention on the basis of packaging design, jingles and gifts. At the same time, but far from view, managers, brokers and investors produce, collect, disseminate and traffic information, not always truthfully, seeking to maximise private gain even at the expense of social losses. Laws and ethical standards are regularly stretched, bent and broken in order to facilitate business transactions, increase market share, extract labour from the workers and draw money from the consumers. Frequent examples of corporate crimes, from the traumatic South Sea bubble of 1720 to the gigantic Enron scandal of 2002, provide a glimpse of the true nature of the 'free market'.[4]

VALUE AND SURPLUS VALUE

The capitalists combine the means of production, generally purchased from other capitalists, with the labour of wage workers hired on the market in order to produce commodities for sale at a profit. The circuit of industrial capital captures the essential aspects of factory production, farm labour, office work and other forms of capitalist production. It can be represented as follows:

$$M - C <^{MP}_{LP} ... P ... C' - M'$$

The circuit starts when the capitalist advances money (M) to purchase two types of commodities (C), means of production (MP) and labour power (LP). During production (... P ...) the workers transform the means of production into new commodities (C'), that are sold for more money (M').

Marx calls the difference between M' and M *surplus value*. Surplus value is the source of industrial and commercial profit and other forms of profit, for example, interest and rent. We are now going to identify the source of surplus value.

Surplus value cannot arise purely out of exchange. Although some can profit from the sale of commodities above their value (unequal exchange), for example unscrupulous traders and speculators, this is not possible for every seller for two reasons. First, the sellers are also buyers. If every seller surcharged his customers by 10 per cent, say, his gains would be lost to his own suppliers, and no one would profit from this exercise. Therefore, although some can become rich by robbing or outwitting others, this is not possible for society as a whole, and unequal exchanges cannot provide a general explanation of profit ('cheating' only transfers value; it does not create new value). Second, competition tends to increase supply in any sector offering exceptional profits, eventually eliminating the advantages of individual luck or cunning (see Chapter 4). Therefore, surplus value (or profit in general) must be explained for society as a whole, or systemically, rather than relying on individual merit or expertise.

A convincing explanation of surplus value and profits must depart from the completely general assumption of equal exchange. Inspection of the circuit of capital shows that surplus value is the difference between the value of the output, C', and the value of the inputs, MP and LP. Since this difference cannot be due to unequal exchange, the value increment must derive from the process of production. More specifically, for Marx, it arises from the consumption of a commodity whose *use value* is to *create new value*.

Let us start from the means of production (physical inputs). In a chocolate factory, for example, cocoa, milk, sugar, electricity, machines and the other inputs are physically transformed into chocolate bars. However, on their own, these inputs do not create new value. The presumption that the transformation of things into other things produces value, regardless of context or human intervention, confuses the two aspects of the commodity, use value and exchange value. It ultimately implies that an apple tree, when it produces apples from soil, sunlight and water, creates not only the use value but also the value of the apples, and that ageing spontaneously adds value (rather than merely use value) to wine. The naturalisation of value relations begs the question of why commodities have value, whereas many products of nature, goods and services have no economic value: sunlight, air, access to public beaches and parks, favours exchanged between friends and so on.

Value is not a product of nature or a substance physically embodied in the commodities. *Value is a social relation between commodity producers that appears as exchange value, a relationship between things* (specifically, value appears through commodity prices, that is, through the relationship between goods and money). Goods and services possess value only under certain social and historical circumstances. The value relation develops fully only in capitalism, in tandem with the production of commodities, the use of money, the diffusion of wage labour, and the generalisation of market-related property rights. At this stage, *value incorporates the most important economic relationships.* Among other things, value relations regulate economic activity, constrain the structure of output and employment, and set limits to social welfare.

If value is a social relation typical of commodity societies, its source – and the origin of surplus value – must be the performance of commodity-producing labour (the productive consumption of the commodity labour power) rather than the metamorphosis of things. When a capitalist hires workers to produce chocolate, for example, their labour transforms the inputs into the output. Because the inputs are physically blended into the output, their value is transferred, and forms part of the output value. In addition to the transfer of the value of the inputs, labour simultaneously *adds* new value to the product. In other words, whereas the means of production contribute value because of the labour time necessary elsewhere to produce them as commodities, newly performed labour contributes new value to the output (see 'Labour' above).

The value of the output is equal to the value of the inputs (MP) plus the value added by the workers during production. Since the value of the means of production is merely transferred, production is profitable only if the value added exceeds the wage costs. In other words, *surplus value is the difference between the value added by the workers and the value of labour power.* Alternatively, *the wage workers work for longer than the time it takes to produce the goods that they command or control.* In the rest of the time, the workers are *exploited* – they produce value for the capitalists. For example, if the goods necessary to reproduce the workforce can be produced in four hours, but the working day is eight hours, the workers work 'for themselves' half the time, and in the other half they work 'for the capitalists': the rate of exploitation (the ratio between what Marx calls 'surplus' and 'necessary' labour time) is 100 per cent.

Just as the workers have little choice on the matter of being exploited, the capitalists cannot avoid exploiting the workers. Exploitation through the extraction of surplus value is a systemic feature of capitalism: this system of production operates like a pump for the extraction of surplus value. The capitalists *must* exploit their workers if they are to remain in business; the workers *must* concur in order to satisfy their immediate needs; and exploitation is the fuel that moves capitalist production and exchange. Without surplus value there would be no wage employment, no capitalist production, and the system would grind to a halt.

It is important to note that, although the wage workers are exploited, they need not be poor in absolute terms (*relative* poverty, due to the unequal distribution of income and wealth, is a completely different matter). The development of technology increases the productivity of labour, and it potentially allows even the poorest members of society to enjoy relatively comfortable lifestyles, however high the rate of exploitation may be. Specifically, if the productivity of labour rises faster than the wage rate (see 'Profit and Exploitation' below), relatively well-paid workers in highly productive economies may be *more* heavily exploited than badly paid workers in less productive economies.

COMPETITION

Competition plays an essential role in capitalist societies. Two types of competition should be distinguished, between capitals in the same sector (producing identical goods) and between capitals in different sectors (producing distinct goods). Firms in the same sector struggle for profits primarily through the introduction of cost-cutting technical innovations. If an innovating firm can produce at a lower cost than its competitors, and they sell at the same price, the more productive firm reaps a higher profit rate and it can increase its market share, invest more and, potentially, destroy the competition. Competition between firms producing similar goods with distinct technologies leads to the *differentiation* of the profit rates (see Chapter 4). This type of competition explains the tendency towards continuous technical progress in capitalism, which is absent in pre-capitalist societies, and it raises the possibility of monopoly and crises of disproportion and overproduction (see Chapter 15).

Competition between firms in distinct sectors is completely different: it generates a tendency towards the *equalisation* of profit rates across the (international) economy. This type of competition explains the equilibrium structures and processes associated with competitive markets, for example, supply adjustments within each sector and capital migration. For example, faced with exceptionally high profits in the Swiss pharmaceutical sector and low profits in the US steel industry, capitalists may decide to invest and thereby increase supply in the former (which eventually lowers pharmaceuticals prices and profit rates), decrease supply in the latter (which eventually raises steel prices and profit rates), migrate from the latter to the former, or pursue a combination of these strategies. What these alternatives have in common is this: they create a tendency towards the equalisation of profit rates across the economy. Inter-sectoral competition, and the tendency towards the equalisation of profit rates, is enormously facilitated by the development of the financial markets.

Capitalist competition has three important implications (explained in more detail in the references listed in note 2). First, it would be misguided to seek an arithmetic solution to the conflicting forces of competition. There is no reason why profit rates should either converge towards an average (which may itself be rising, falling or static), or diverge permanently, potentially leading to the development of super-monopolies. The two types of competition explained above influence the behaviour of firms in different ways, and the outcome of their interaction (and other influences on firms' behaviour) depends upon a wide range of variables that can be understood only concretely (see Chapter 16). Second, price changes due to inter-sectoral competition influence the operation of the law of value. Rather than commodity exchanges being regulated simply by the abstract labour time necessary to produce commodities, as in Smith's rude society, in advanced capitalism prices depend upon the equalisation of profit rates between sectors of the economy (this is known as the 'transformation of values into prices of production'; see Chapter 4). Third, the interplay of the forces of competition within and between sectors generates a tendency towards the reduction of the quantity of labour required in production across the economy (this is known as the 'tendency for the rate of profit to fall', which Marx analysed simultaneously with the 'counter-tendencies' to this law; see Chapter 15).

PROFIT AND EXPLOITATION

The profits of firms can increase in many different ways. For example, the capitalists can compel their workers to work longer hours or work harder (greater intensity of labour), employ better skilled workers, or change the technology of production.

All else being constant, longer working days produce more profit because more output is possible at little extra cost (the land, buildings, machines and management structures being the same). This is why capitalists always claim that the reduction of the working week would hurt profits and, therefore, output and employment. However, in reality other things are not constant, and historical experience shows that such reductions can be neutral or even lead to higher productivity because of their effects on worker efficiency and morale. Outcomes vary depending on the circumstances, and they may be strongly negative for some capitalists and, simultaneously, highly profitable for others.

Greater labour intensity condenses more labour into the same working time. Increasing worker effort, speed and concentration raises the level of output and reduces unit costs; therefore, profitability rises. The employment of better trained and educated workers leads to similar outcomes. Such workers can produce more commodities, and create more value, per hour of labour.

Marx calls the additional surplus value extracted through longer hours, more intense labour or the employment of better trained workers *absolute surplus value*. This type of surplus value involves the expenditure of more labour, whether in the same working day or in a longer day, with given wages. Absolute surplus value was especially important in early capitalism, when the working day was often stretched as long as 12, 14 or even 16 hours. More recently, absolute surplus value has often been extracted through the lengthening of the working week and the penetration of work into leisure time, at least for certain sectors of the workforce (work often extends into the weekend and holidays, and the general availability of mobile phones and portable computers allows the employees to be permanently on duty). Moreover, the workers are frequently compelled to increase productivity through more intense labour (for example, faster production lines or reduced breaks) or coerced into acquiring new skills in their 'free' time (for example, by attending conferences and courses). In spite of its importance, absolute surplus value is limited. It is impossible to increase the working day or the intensity of labour indefinitely, and the workers gradually learn to resist these forms of exploitation.

The introduction of new technology and new machines can also increase the profit rate of the innovating firms. They allow more inputs to be worked up into outputs in a given labour time or, in other words, they reduce the quantity of labour necessary to produce each unit of the product. When productivity rises faster than wages across the economy, the share of surplus value in the total value added increases and the workers' share declines. Marx calls this *relative surplus value*. Relative surplus value is more flexible than absolute surplus value, and it has become the most important form of exploitation under modern capitalism, because productivity growth can outstrip wage increases for long periods (the implications of absolute and relative surplus value are discussed in Chapter 5, and the use of new technology in order to control the workforce is analysed in Chapter 6; see also Saad-Filho 2002, ch. 5).

OVERVIEW AND CONCLUSION

Mainstream economic theory defines capital as an ensemble of things, including means of production, money and financial assets. More recently, human knowledge and community relations have been named human or social capital. This is incorrect. These objects, assets and human attributes have existed for a long time, whereas capital is relatively recent. It is misleading to extend the concept of capital where it does not belong, as if it were valid universally or throughout history. For example, a horse, a hammer or a million dollars may or may not be capital; that depends on the context in which they are used. If they are engaged in production for profit through the direct or indirect employment of wage labour, they are capital; otherwise, they are simply animals, tools or banknotes.

Like value, *capital is a social relation that appears as things*. However, whereas value is a general relationship between the producers and sellers of commodities, *capital is a class relation of exploitation*. This social relationship includes two classes (defined by their ownership, control and use of the means of production): the capitalists, who own the means of production, labour power and the product of labour, and the wage workers, who sell their labour power and operate the means of production without owning them. The relationship between these two classes is the basis for the social division of labour and the production and distribution of commodities.

Competition and exploitation through the extraction of surplus value render capitalism uniquely able to develop technology and the forces of production. This is the main reason why Marx admires the progressive features of capitalism. However, capitalism is also the most *destructive* mode of production in history. The profit motive is blind, and it can be overwhelming. It has led to astonishing discoveries and unsurpassed improvements in living standards, especially (but not exclusively) in the developed countries. In spite of this, capitalism has also led to widespread destruction and degradation of the environment and of human lives. Profit-seeking has led to slavery, mass murder and even genocide (for example, against the native populations of the Belgian Congo and the United States, in South Africa under apartheid and in colonial and inter-imperialist wars, most clearly in the First World War), brutal exploitation of the workers (in nineteenth-century Britain, twentieth-century Brazil and twenty-first-century China), and the uncontrolled destruction of the environment (in the United States, Europe, India, Indonesia and elsewhere), with long-term global implications (see Chapter 7).

Capitalism both generates and condones the mass unemployment of workers, machinery and land in spite of unsatisfied wants, and tolerates poverty even though the means to abolish it are readily available. Capitalism extends the human life span, but it often empties life of meaning. It supports unparalleled achievements in human education and culture while, simultaneously, fostering idiocy, greed, mendacity, sexual and racial discrimination and other forms of human degradation. Paradoxically, the accumulation of material wealth often impoverishes human existence.

These contradictory effects of capitalism are inseparable. It is impossible to pick and choose the appealing features of the 'market economies' and discard those that we find distasteful. Private ownership of the means of production and market competition *necessarily* give rise to the wage relation and to exploitation through the extraction of surplus value, and they facilitate crises, war and other negative features of capitalism. This places a strict limit on the possibility of social, political and economic reforms, and on the capacity of the market to assume a 'human face'.[5]

Limitations such as these led Marx to conclude that capitalism can be overthrown, and another social system created, communism. For him, communism opens the possibility of realisation of the potential of the vast majority through the elimination of the irrationalities and human costs of capitalism, including systemic inequality, material deprivation, destructive competition, greed and economic exploitation (this system, and the transition towards it, are discussed in Chapters 18 and 19).

REFERENCES AND FURTHER READING

Fine, B. (1989) *Marx's Capital* (3rd edn.). Basingstoke: Macmillan.
Fine, B. (2002) *The World of Consumption*, (2nd edn.). London: Routledge.
Foley, D. (1986) *Understanding Capital, Marx's Economic Theory*. Cambridge, Mass.: Harvard University Press.
Harvey, D. (1999) *The Limits to Capital*. London: Verso.
Marx, K. (1976, 1978b, 1981) *Capital* (3 vols.). Harmondsworth: Penguin.
Marx, K. (1978a, 1969, 1972) *Theories of Surplus Value* (3 vols.). London: Lawrence and Wishart.
Marx, K. (1981) *Grundrisse*. Harmondsworth: Penguin.

Marx, K. (1987) *A Contribution to the Critique of Political Economy* (Collected Works, vol. 29). London: Lawrence and Wishart.

Perelman, M. (2000) *Transcending the Economy: On the Potential of Passionate Labor and the Wastes of the Market.* New York: St. Martin's Press.

Saad-Filho, A. (2002) *The Value of Marx: Political Economy for Contemporary Capitalism.* London: Routledge.

Smith, A. (1991) *Inquiry into the Nature and Causes of the Wealth of Nations.* London: Everyman.

Weeks, J. (1981) *Capital and Exploitation.* Princeton: Princeton University Press.

Wood, E.M. (1988) 'Capitalism and Human Emancipation', *New Left Review* 167, pp. 3–20.

Wood, E.M. (1999) *The Origin of Capitalism.* New York: Monthly Review Press.

NOTES

1. I am grateful to Andrew Brown, Paul Burkett, Ben Fine, Costas Lapavitsas, Simon Mohun and Alejandro Ramos for their valuable comments on previous drafts of this chapter.
2. For overviews of Marxian value theory at different levels of difficulty, see Fine (1989), Foley (1986), Harvey (1999), Saad-Filho (2002) and Weeks (1981).
3. Systems of provision are discussed in detail by Fine (2002).
4. For an outstanding study of the wastes of the market, see Perelman (2000).
5. See Wood (1999).

2   Does All Labour Create Value?

Simon Mohun

HISTORICAL AND INTELLECTUAL ORIGINS

The industrialisation of Britain from the middle of the eighteenth century transformed both rural and urban environments. Manufacturing in the cottages of the countryside (the 'putting-out system') was gradually centralised in larger units (the 'factory' system) typically located in the rapidly growing towns. This enabled substantial economies of scale through the use of newly harnessed sources of power, the further development of the division of labour, and the much closer control that could be exercised over the production process. At the same time, agricultural enclosures of common land dispossessed the rural poor of their traditional grazing and foraging rights (see Chapter 8). The combination of the decline of cottage industry with the enclosures of common land deprived large numbers of rural families of their livelihood. The complex and precarious ways in which a rural family survived, through a combination of agricultural wage labour, cottage industry, family labour on a smallholding and access to common land, was increasingly attenuated, and more and more families were compelled to seek subsistence entirely through the market. Typically, the only commodity they had to sell was their own capacity to work. Only the sale of this capacity (their labour power) for a wage could provide them with the money required for the purchase of the commodities necessary for subsistence (see Chapter 1). In this manner, a landless working class was created and industrialisation proceeded, increasingly an urban phenomenon.

In the late eighteenth century, contemporaries were aware of the beginnings of these processes, in terms of both their novelty and their scale, and attempted to theorise the phenomena they were witnessing. Adam Smith's *Inquiry into the Nature and Causes of the Wealth of Nations* (1776) focused on the benefits from specialisation as the division of labour was extended. He saw these benefits as limited only by the extent of the market. Indeed, he linked the extension of the division of labour with the extension of the market in a mutually reinforcing process: specialisation increases productivity and incomes; this stimulates both investment and demand (and after a lag, population growth), which widens the market; and this in turn enables further specialisation. The role of government was limited to encouraging these processes by guaranteeing internal and external security (for both people and property), and maintaining a legal system and a stable currency.

However, an important question that worried Smith was whether all employment contributed to this virtuous growth cycle. This concern did not originate with Smith. In the 1690s Gregory King attempted a statistical description of English society for the year 1688, in which more than half of the population was categorised as 'Decreasing the Wealth of the Kingdom', meaning dependent to some degree on transfer payments.[1] And within the developing discipline of political economy, the sources of wealth tended to be located in the activity of some particular sector (for the mercantilists, the acquisition of bullion through foreign trade; for the physiocrats, an agricultural surplus), thereby defining economic activity in other sectors as unproductive. So this was an important issue for Smith to confront.

Smith took a broader view than earlier writers, and designated as productive the labour that contributed to a positive feedback between extension of the division of labour and growth of the market. Employment of such labour was effectively an investment, contributing more to output than it cost in wages. Otherwise, labour was unproductive, contributing nothing to the growth of output by its activity, and consuming a portion of total output by virtue of the wages it cost. An example of productive labour might be a worker in one of the new cotton mills. She is paid a wage and is part of a division of labour that produces an output that is sold, from the proceeds of which the capitalist recovers his outlay of wages and gains a profit that provides the funds for further investment. An example of unproductive labour might be a worker in domestic service. She is paid a wage (partly in cash, partly in kind) in return for an output (domestic service) that is not sold on the market but is directly consumed by her employer. Payments to such a worker are a net cost to the economy.[2]

But Smith's attempt to draw a clear line of demarcation between productive and unproductive labour in the terms just outlined is seriously confused by a different distinction he draws, in which productive labour produces a physical product, and unproductive labour produces a service. It is easy to see how this second definition arises, because Smith wanted to contrast the growing and productive manufacturing sector, which typically produces a physical output, with the small armies of retainers unproductively employed in service by the landed gentry, which he saw as consuming rather than producing output.[3] In an economy in which marketed services are negligible, the two lines of demarcation are very similar. But as soon as services are marketed to a significant extent, the two definitions are incompatible. And there is a further confusion, to do with the contrast between producing and consuming output. For an activity might be profitable for an individual employer, and yet add nothing to social output, so that what is productive from a private perspective might be unproductive from the perspective of society. If for example the profits on some (unproductive) activity were in fact a market transfer out of the profits of some (productive) activity, the unproductive activity would appear productive when considered in isolation, and yet contribute nothing to aggregate profits and hence be unproductive when the economy as a whole is considered.

In the early stages of the industrial revolution, it was perhaps inevitable that these inconsistencies were not so obvious. But by the middle of the nineteenth century, Smith's definitions were an increasingly unreliable guide. Their interest is that they provided the starting point for Marx's analysis of productive and unproductive labour.

PRODUCTIVE AND UNPRODUCTIVE LABOUR

Marx absorbed Smith's vision of a dynamically growing economy and developed further Smith's first distinction between productive and unproductive labour, but within a rather different framework. First of all, and obviously, in any society, labour that produces anything useful is productive. The difficulty is that what is regarded as useful is historically specific, and is conditioned and structured by the framework set by the dominance of some particular relations of production. It is therefore first necessary to consider those class relations directly. What differentiates class societies is the form in which the dominant class is able to extract surplus labour from the subordinate class. In capitalist society, surplus labour takes the form of a sum of money, called surplus value or profit. Accordingly for Marx, any labour in capitalist society is productive if and only if it produces surplus value.

Several points should be noted about this definition. First, the nature of the output (for example, whether a physical good or an intangible service) is irrelevant. Only the social relations under which it is produced count. Hence a necessary condition for labour to be productive is that it is *wage* labour. Secondly, since wage labour must produce surplus value, or profit, to be productive, and profit only derives from the sale of output, a further necessary condition for labour to be productive is that the output it produces is *marketed*. Thirdly, the activity in which productive labour is engaged is a transformative activity of *production*. The activity cannot be one which distributes or redistributes an output which has already been produced elsewhere, and nor can it be one whose function is to collect together inputs so that they are then ready for production. These types of activity earn profit that is a redistribution (through the market via the price mechanism) of total profits deriving from production, and so do not contribute in the aggregate to total profits produced. Hence a further necessary condition for labour to be productive is that *additional* surplus value is produced. In sum, in capitalist society, productive labour *first*, is wage labour, *second*, is employed in a capitalist production process, and *third*, produces surplus value from a social point of view. All other wage labour is unproductive.

The implications of each of these necessary conditions are important. The first condition requires labour to be wage labour if it is to count as productive. Labour that is not wage labour is not productive. That this says nothing about the *necessity* for such non-wage labour can be seen from the fact that in any society an enormous amount of time is spent in informal and unwaged caring activities, looking after the young and the old. No society could reproduce itself without at least the labour time spent in creating and caring for children, but all workers engaged in such unpaid caring activities are unproductive. They produce neither value nor surplus value; for all that their work is essential.

Secondly, not all wage labour is productive. Output has to be sold in order that surplus value be appropriated; hence output produced by wage labour that is not marketed cannot produce any surplus value. In any society, substantial numbers are employed in a wage labour relation by 'general government'. General government produces output for individual and/or collective consumption that is consumed directly, makes cash transfers, and invests in public assets. Its activities are financed by levying taxes and selling financial instruments.[4] General government activities include general public services (executive, legislative and judicial), internal (police) and external (armed services) security, welfare services (health, education, social security, housing) and economic services (administration of subsidies and other interventions in industry). Hence general government employs a substantial number of people, but none of them produces either value or surplus value, and hence they are all unproductive.

Thirdly, whether wage labour produces surplus value can be determined only from an overall social perspective. For capitalist employment of wage labour producing a marketed output and earning profits might nevertheless consume rather than add to total surplus value. Consider for example workers employed by a profitable advertising agency. The agency is contracted by a firm to run a campaign on the firm's commodity. The only output (if the campaign is successful) is increased sales of the firm's commodity, and, whether successful or not, the agency is paid out of the revenues accruing from the firm's sales. The agency therefore produces nothing, and is paid out of a transfer of resource from the contracting firm. No matter that the advertising agency might persuasively create demand and thereby extend the market; what it does is to facilitate the sale of commodities produced elsewhere. Generalising from this example, all labour that is employed one way or another purely to sell output is involved in facilitating a transfer of title of ownership. Since nothing additional is produced by that labour, then that labour is not productive. The surplus value deriving from such commercial activities arises not from the exploitation of workers employed in those activities, but from a transfer through the price mechanism of profit produced by productive workers elsewhere. Whereas the capital that employs workers who produce surplus value is called 'industrial capital', the capital that employs workers to buy and sell the products of industrial capital is called 'commercial capital'. Commercial capital appropriates a portion of the surplus value produced by industrial capital via an unequal exchange. The more sophisticated is the knowledge required about the market, the more commercial capital can carve a specialised niche for itself.

Symmetrically, the same point can be made about all of those activities that facilitate the purchase of inputs. Large numbers of people are employed in these activities, typically involved in recording and accounting for financial flows, and transferring title to sums of money and to increasingly complicated financial instruments representing sums of money. The capital that employs workers in these sorts of activities is called 'financial capital'. The functions of financial capital are in general to organise and operate in financial markets, to spread risk, to consolidate smaller sums of money into larger ones, and to provide credit. In this manner, large sums of capital are made available for the purchase of inputs by industrial capital (see Chapters 3 and 10). The typical payment is a rate of interest, which determines a transfer of value between the two contracting parties. But despite the commodity form of a financial service, there is no commodity produced, hence no commodity equivalent to match the payments of interest. Consequently, interest payments must be understood in terms of exploitation and unequal exchange. Like the net earnings of commercial capital, interest payments are in general a claim on the surplus value produced by industrial capital. The only difference is that the activities of commercial capital realise surplus value that has already been produced, whereas the activities of financial capital are paid for out of a pre-commitment by industrial capital of surplus value yet to be produced. Hence in this latter case, a speculative element is involved. Figure 2.1 summarises this vision of the capitalist production process.

As soon as financial capital is used to purchase inputs for production, that capital, as an amount of value, changes its form from financial to productive capital. Despite the change in form, the quantity of value does not change. Once inputs have been consumed in the production process to produce output, the capital becomes commercial capital (called 'commodity capital' by Marx). Now its quantitative value *has* increased, by virtue of the difference between what labour power cost and what labour can produce. When the output is sold, the sum of gross value produced takes a financial form, to be recommitted to the production process in due course. Again, in this change of form, the quantity of value does not change. The only quantitative change in value (an expansion) takes place in production, following the advance of capital to purchase inputs and prior to the appearance of commodity outputs and their sale. Selling the output, operating in money markets and purchasing inputs all transfer the form in which value exists, but they do not alter its quantity.

Figure 2.1 is a highly stylised and abstract representation. The activities of both commercial capital and financial capital can in practice be very complex, but analytically the surplus value that they earn remains a transfer from the surplus value deriving from production. The labour power hired by commercial and financial capital is exploited, like any other labour power, if workers are

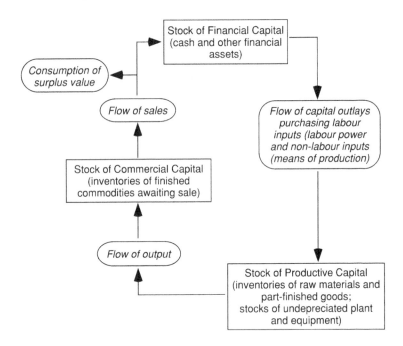

Figure 2.1   The Circuit of Capital

Source: adapted from Foley (1986), pp. 66–9.

compelled to work for longer than is required to produce their consumption needs. But this unpaid labour time is not monetised; the surplus value accruing to commercial and financial capital derives from a transfer through the market of surplus value produced by industrial capital. This occurs through interest payments and fees charged by financial capital, and through fees charged by commercial capital combined with unequal exchange (for processes of pure sale, such as commodity broking). In sum, transformations of form, from commodity output into money, from money into other financial assets and back into money, and from money into commodity inputs, do not change the quantity of value that exists.

That 'productive' strictly means 'productive of surplus value' means that the theoretical term has no trans-historical meaning. It is only concerned with what is productive, what is to count as social productivity, under specifically capitalist relations of production. One might be morally offended that the surgeons, nurses and technicians engaged in vital organ transplants in a state hospital are unproductive labour, whereas private sector workers producing weapons designed to destroy such organs are productive labour. But that is to be offended at the prevailing relations of production, in which production is organised by considerations of private profit rather than considerations of social need. For as long as production is so organised, the class criterion is paramount: labour is productive if it produces surplus value.

CONTROVERSIES

Like other Marxian categories, the categories of productive and unproductive labour have no counterpart in the different theoretical framework of neoclassical economics. For the latter, anything whose consumption contributes to someone's utility can command a price in the market and return a revenue stream to its owner, and so corresponds to the production of a good or service.[5] Categories of productive and unproductive labour are therefore meaningless: in general *any* good or service supplied is the outcome of a production process, and its price (whether real, potential or shadow) is a return to its owner. Echoes of the distinction between productive and unproductive labour sometimes surface in concerns about the size of the state sector and its effects on growth. But it is not that the state sector is 'unproductive' in neoclassical economics. It is rather that since the state sector is financed by compulsory taxation, too large a state sector requires levels of taxation which will generate disincentive effects at the margin in the private sector on both labour–leisure tradeoffs and the investment decision. In sum, for the neoclassical tradition, notions of productive and unproductive labour simply make no sense.

Matters are different for a labour theory of value. But even within this tradition, there is considerable controversy about whether the distinction between productive and unproductive labour is tenable.[6] These controversies can be summarised in terms of each of the three points emphasised above: first, that productive labour has to be wage labour, second, that it has to produce a marketed output, and third, that it is engaged in production.

One line of questioning has been to refocus the meaning of 'productive' as necessary or essential. To define some activity as unproductive carries the connotation that it is unnecessary, and this slights or denigrates the people engaged in such activity. Consider unpaid housework and childcare. These are activities that are predominantly undertaken by women, and to call such activities unproductive has been interpreted as an example of the way in which a patriarchal theory systematically ignores the activity of women. A response might be that this confuses the reproduction of capitalist relations of production with the reproduction of the wider society. There are all sorts of complex interrelations between the two, structured by the evident truth that children (and thus future workers) are not produced under capitalist relations of production. Therefore the reproduction of those relations requires a permanent flow of inputs of labour power 'produced' (at least in part) elsewhere, in the family and the school. But the reproduction of capitalist relations and the reproduction of wider society are not identical, and the theoretical categories used to analyse the one are inappropriate for the other.

Secondly, consider wage labour that does not produce a marketed output, typically employed by general government. General government in 2000 accounted for about 13.4 per cent of all employment in the UK, and for about 12.9 per cent in the USA.[7] So the numbers involved in developed capitalist economies are substantial. The questioning of the productive–unproductive distinction here also focuses on a denial of the distinction between 'necessary' and 'productive'. If society cannot function without general government, then it makes little analytical sense to call general government employees unproductive, and indeed concedes too much to pro-market ideology. But the response is the same as that already given: it is important not to confuse the reproduction of capitalist relations of production with the reproduction of the wider society. A more specific but related line of questioning concerns the activities involved in the maintenance and training of the working class. If extra skills are acquired by a worker through consuming some state sector output of education and training, then that worker will produce more value in a given time period than an untrained but otherwise identical worker. Similarly, a healthier worker will have lower maintenance and reproduction costs than a less healthy one. In this manner, state provision of education, training and health contribute to the production of surplus value, not directly, but indirectly through transformations of the quality of the living labour input into the capitalist production process. Then it is argued that there is no reason to separate those activities that are directly productive of surplus value from those that are indirectly productive. The difficulty with this argument is its very breadth. If all activities that indirectly contribute to surplus value are considered productive, the term loses any focus and precision, for it is hard to conceive of any activity that cannot be so interpreted.

A third line of questioning, perhaps the most influential, has been to focus on what 'production' means.[8] In particular, it is argued that it is not possible to make a hard distinction between production activities, in which inputs are combined in a production process organised by industrial capital to produce an output, and circulation activities, in which outputs are transformed into money that is then reinvested in inputs by the activities of commercial and financial capital. There are only two ways in which critics have argued that such a separation can be conceived. One way is by reverting to Smith's second definition, in which labour is productive if it produces a physical good, for only a resort to 'physicalism' can adequately determine what is produced from what is circulated. The other way is to define as unproductive what is specific to capitalism, by reference to an evaluative standpoint based on communism. For example, if communist distribution is direct rather than through the market, then the labour involved in marketing activities will not exist under communism, and is therefore unproductive in capitalist society. Since communist production is for need rather than for profit, there will be no advertising, and so advertising labour is unproductive, and so on. To identify unproductive labour on this evaluative criterion is to locate sources of waste in contemporary capitalist society, and to identify resources that a more progressive society can employ to increase the production of use values for the benefit of all.

The 'physicalist' criterion bears no relation to capitalist social relations, and is not therefore a helpful one for the analysis of contemporary capitalism. The 'evaluative' criterion, while perhaps determining a useful project in the identification of waste, also bears no relation to the analytical categories of the labour theory of value, and so again is not helpful in the present context. But it is argued that, unless they resort to one or other of these criteria, all attempts to found a distinction between productive and unproductive labour fail. The distinction is empty, and should be abandoned. The circuit of capital should be understood as a metaphorical rather than a literal description of how surplus value is produced and realised. To separate production from circulation, with productive labour confined to the former and unproductive labour to the latter, is to separate in an artificial and mechanistic way what are distinct yet simultaneous components of the same social process.

The response to this line of questioning has been to deny that the distinction is analytically empty, and to assert on the contrary that it is fruitful at both theoretical and empirical levels. The distinction between industrial capital, on the one hand, and commercial and financial capital, on the other, enables a grasp of the changing organisation of capitalism as their autonomy from each other develops alongside their dependence on each other, a continually fluctuating balance of power now favouring the one, now the other. Focusing on the development of unequal exchange and the dependence of commercial and financial profits on the surplus value produced by industrial capital is to focus on both the possibilities and the limits of specialisation by capital in particular historical periods. A labour theory of value which includes the categories of productive and unproductive labour yields a richer picture of capitalist development, and one that is more consonant with what one would expect to be shown by Marxian theory, than a labour theory of value that abolishes the distinction.

USES OF THE DISTINCTION

The productive–unproductive labour distinction is important in analysing the development and relative strengths of fractions of capital (and indeed alliances cutting across those fractions). It focuses attention upon the dependence of other fractions upon industrial capital, and hence enables investigation of why that dependence might be tighter or looser in particular periods. It also differentiates the determinants of sectoral profitability and the ways in which different capitals participate in the competition which tendentially results in all capitals earning the same rate of profit.

Whereas for industrial capital, profitability is determined by the productivity of labour, the organisation of the labour process, and the level of wages, for commercial capital, given some cost structure, profitability is determined by its ability to charge fees for its services and to increase unequal exchange. These both depend upon its position in the market, its degree of specialised commercial knowledge, and its ability to organise networks of distribution. Competition between commercial capitals will tend to reduce unequal exchange to the level at which revenues are sufficient for each commercial capital to earn the average rate of profit. For financial capital, given some cost structure, profitability is determined by the difference between borrowing and lending rates and its ability to charge fees for money market operations. In a world of certainty, arbitrage between capitals should ensure equality between the interest rate and the average rate of profit, with the difference between borrowing and lending rates just sufficient to cover the costs of making loans. But real uncertainties in the production and realisation of surplus value by industrial capitals can serve to differentiate the interest rate from the profit rate, which throws the focus on to the ways in which levels of future profitability of industrial capitals affect the determination of the current rate of interest.

Secondly, the productive–unproductive labour distinction has some importance in analysing the changing historical determinants of general government expenditure. In general terms, Marx saw the state as representing the interests of capital as a whole. At the same time the state has in some sense to manage class conflict. On the one hand, one consequence of working-class struggle is that some activities are taken out of private production and into state collective provision financed by taxation. On the other hand, the state attempts to organise the provision of its activities in ways most beneficial to capital. The changing balance of class forces at any time shapes how these factors historically combine. Compare for example the decade after 1945 in Western Europe with its nationalisations and other forms of state regulation together with the development of state education, health and social insurance, with the 1980s and 1990s and their privatisations together with tight restrictions on expenditure on state education, health and social insurance.

The third way in which the distinction is of some use is in empirical investigations of capitalist development. Given the nature of the concepts and the data available, too much precision should not be expected. In particular, data are generally organised by an industrial classification, and assumptions must always be made about how to divide productive from unproductive both across industrial classifications and within them. Because there are always borderline cases, and because of data limitations, occasionally arbitrary and sometimes heroic assumptions must be made. But while exact precision is impossible, some reasonable estimates of time trends are possible.

Consider for example the UK in the census years 1861 to 1911 as shown in Table 2.1. An example of arbitrariness is the allocation of all services except domestic service to productive labour, whereas some will certainly be unproductive. An example of approximation is the determination of productive labour in each productive sector by the proportion of wages to salaries in that sector, roughly

Table 2.1   Productive and Unproductive Labour by Industry (thousands), UK, Census Years

	1861	1871	1881	1891	1901	1911
Agriculture and fishing	3017	2663	2369	2181	1924	1820
Mining and quarrying	420	486	563	697	811	978
Manufacturing	3686	4011	4075	4578	4762	4967
Building and contracting	471	563	687	697	867	781
Gas, electricity, water	21	26	33	50	79	91
Transport and communication	506	649	712	921	1153	1198
Health, education and other professional services	287	333	439	498	572	629
Catering, hotels and other services	386	444	555	680	708	804
Total productive	8794	9174	9433	10300	10875	11268
Productive as % of total	67.2	65.3	62.6	61.8	58.2	55.3
Distributive trades	850	1050	1300	1640	1990	2460
Insurance, banking, finance	20	40	70	110	150	230
Public administration and defence	450	420	460	550	880	840
Private domestic service	1294	1790	1850	1940	1980	2000
Unproductive labour in productive sectors	1681	1576	1957	2120	2805	3592
Total unproductive	4296	4876	5637	6360	7805	9122
Unproductive as % of total	32.8	34.7	37.4	38.2	41.8	44.7
Ratio of Unproductive to Productive Labour	0.49	0.53	0.60	0.62	0.72	0.81

Source: derived from Feinstein (1976) Tables 60 and 21

measuring a 'blue collar' (factory operative) 'white collar' (office worker) distinction, which is not exactly what is required. This notwithstanding, the upward drift in unproductive labour is striking. Comparing 1861 with 1911, the main movements are the fall in productive labour in agriculture from 23.0 per cent to 8.9 per cent of total employment; the rise in distributive trades employment from 6.5 per cent to 12.1 per cent of total employment, and the rise in unproductive labour in productive sectors from 12.8 per cent to 17.6 per cent of total employment. These years of the 'second industrial revolution' see a growth in specialised marketing activities, but with scope for much further specialisation by productive firms in outsourcing their growing unproductive activities.

Another question to consider is how unproductive labour has affected the general rate of profit, although this is not a simple

question (see Chapters 15 and 16). Define the pre-tax rate of profit
(r) as the ratio of aggregate profits to the aggregate net capital stock
(K), and define profits as the difference between adjusted net
national product ($adj.NNP$) and total private sector wages.[9] These
latter are the wages paid to productive labour (W_P) and the wages
paid to unproductive labour (W_U). Hence:

$$r = \frac{Adj.NNP - W_P - W_U}{K}$$

$$= \frac{\dfrac{Adj.NNP - W_P}{W_P} - \dfrac{W_U}{W_P}}{\dfrac{K}{W_P}}$$

The first term in the numerator is the money form of the rate of
surplus value (e), so that:

$$r = \frac{e - \dfrac{W_U}{W_P}}{\dfrac{K}{W_P}}$$

The ratio of unproductive to productive labour in wage terms is
a direct negative influence on the rate of profit, but might
indirectly positively affect the rate of profit if the specialisation of
function enabled by the contracting out of unproductive activities
by productive capital increases the rate of surplus value. Thus the
ratio of unproductive to productive labour in wage terms is an
important one, and its growth in the USA in the last third of the
twentieth century is shown in Figure 2.2. Since the data is in
natural logs, the slope of the line representing the ratio is the rate
of growth of the ratio.

The data divide into three distinct periods. From 1964 to 1978 the
ratio grows in total by 7.9 per cent, periods of positive growth being
interspersed with two periods of negative growth. This fluctuating
but fairly flat overall period is marked by the Vietnam War, the
collapse of Bretton Woods, and the stagflation of the 1970s. From
1978 to 1992, dominated by the expansion of the Reagan–Bush
years, there is a more sustained increase, in which the ratio grows by

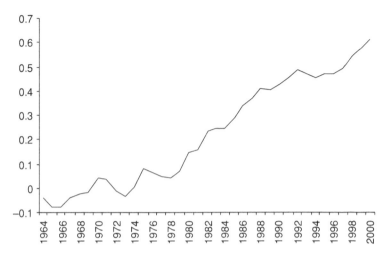

Figure 2.2 Wage Ratio of Unproductive to Productive Labour, USA, 1964–2000 (natural logs)

a total of 44.5 per cent. This is followed by another essentially flat period during the first Clinton presidency (from 1992 to 1997 the ratio falls and then rises, growing by a total of 0.35 per cent). And the twentieth century concludes with what looks like another sustained increase as the ratio grows by 11.9 per cent from 1997 to 2000 (although this may be subject to data revisions). Periods of higher output growth appear to allow significant relative increases in unproductive labour, whereas periods of lower output growth do not (although it remains to be seen whether the increases of the second half of the 1990s will be maintained into the twenty-first century). Combined with an analysis of profitability, and some assessment of the effects of the relative increase of unproductive labour on that profitability, these figures provide a basis for an empirical analysis of structural change in the US economy in the last third of the twentieth century.

REFERENCES AND FURTHER READING

Feinstein, C.H. (1976) *Statistical Tables of National Income, Expenditure and Output of the UK 1855–1965*. Cambridge: Cambridge University Press.

Foley, D.K. (1986) *Understanding Capital*. Cambridge, Mass.: Harvard University Press.

Laibman, D. (1992) *Value, Technical Change and Crisis*. Armonk, New York: M.E. Sharpe.

Laibman, D. (1999) 'Productive and Unproductive Labour: A Comment', *Review of Radical Political Economics* 31 (2), pp. 61–73.

Laslett, P. (2000) *The World We Have Lost Further Explored*. London: Routledge.

Mitchell, B.R. (1988) *British Historical Statistics*. Cambridge: Cambridge University Press.

Mohun, S. (1996) 'Productive and Unproductive Labour in the Labour Theory of Value', *Review of Radical Political Economics* 28 (4), pp. 30–54.

Mohun, S. (2002), 'Productive and Unproductive Labour: A Reply', *Review of Radical Political Economics* 34 (2), pp. 187–201.

Office for National Statistics (2001a) 'Jobs in the Public and Private Sectors', *Economic Trends,* June.

Office for National Statistics (2001b) *UK National Accounts (Blue Book) 2001 Edition*. London: The Stationery Office.

Shaikh, A.M. and Tonak, E.A. (1994) *Measuring the Wealth of Nations*. Cambridge: Cambridge University Press.

Smith, A. (1776) *Inquiry into the Nature and Courses of the Wealth of Nations*. London: Everyman (1991).

US Bureau of Economic Analysis: *National Income and Product Accounts* (www.bea.gov).

US Bureau of Labor Statistics (stats.bls.gov).

NOTES

1. King's table is reproduced and discussed in Laslett (2000), pp. 30ff. King's data are revised by Lindert and Williamson and reproduced in Mitchell (1988) ch. 2, p. 102.
2. Smith did not consider how expenditure out of the wages of unproductive workers adds to overall demand and thereby indirectly contributes to the extension of the market.
3. The number of families in the category 'high titles and gentlemen' in England and Wales was 19,626 in 1688, 18,070 in 1759 and 27,203 in 1801/03. See the sources cited in note 1.
4. Fees might be charged for some portions of general government output, but they are not economically significant in terms of cost recovery of the activities concerned. General government in some very poor countries might also depend upon the receipt of grant aid from overseas.
5. Some qualification is necessary. Sometimes, the market transaction is only a potential one. Thus homeowners are deemed to pay a rent to themselves, which is counted as a return for the production of housing services, for otherwise national income would fall whenever a renter purchases a house. And sometimes markets cannot exist for technical reasons. If consumption of a good by one person does not diminish the amount available to another person, and if nobody can be excluded from consuming the good, then the good is a pure 'public good' and must be financed out of taxation.
6. See Mohun (1996, 2002) and Laibman (1999), and the references therein.
7. The figures are on a 'full-time equivalent' basis, and for the UK include those employed by National Health Service Trusts. The UK figures are from the Office for National Statistics (2001a) and (2001b), and the US figures

are from the US Bureau of Economic Analysis *National Income and Product Accounts*.

8. See Laibman (1992, ch. 4, and 1999).

9. Net national product should be adjusted downwards for three reasons. Imputations should be subtracted, because they correspond to a flow of services which is not marketed; general government wage costs should be excluded, because general government workers are financed out of taxation rather than the market; and household worker wage costs should be subtracted, because no output is sold.

3 Money as Money and Money as Capital in a Capitalist Economy

Costas Lapavitsas

Money permeates economic activity in capitalism, from the mundane to the vital. Money also permeates social life, making or breaking personal relations, attaching meaning to human action and providing its holders with various human qualities. But despite its prominence in capitalism, there is no consensus in social theory on what money is and how it functions. This chapter considers the social relations that give rise to money and those that rest on it, from the perspective of Marxist political economy. The first section (Money as Money) focuses on money as plain money, that is, money as a phenomenon of simple commodity exchange. By considering money purely in the context of market trading, it is possible to specify what money is as well as its functions and forms in relation to markets. The second section (Money as Capital) turns to money as capital, that is, money as a phenomenon of capitalist production and circulation. Money's characteristically capitalist functioning is thus specified, including its role in relation to credit.

MONEY AS MONEY

Money and markets

Capitalism is a social system that incorporates an extremely wide network of markets. There are markets in which the traded commodities are produced by capitalist enterprises employing wage labour, such as those for consumer and investment goods. There are markets in which the traded commodities are not produced by using capitalist methods, typical examples being the markets for land and labour (see Chapters 1 and 4). There are also markets in which the objects of trading are not produced commodities at all, but financial obligations, claims on others, cover for risk and other promises among people. Finally, there are even 'markets' in which the traded objects can only be imputed by analogy with commodity markets, such as the 'markets' for bribes, for gangster protection, for hired murderers, for fines, for libel compensation, and so on. All these disparate markets, however, have one thing in common: money.

The functions of money in these capitalist markets are ubiquitous. Money is the means of rendering disparate objects and activities commensurate with each other (the unit of account or measure of value). It is the mediating instrument in transactions (the means of exchange). It is, further, the medium that enables settlement of promises and obligations between market participants at a time other than that of the actual transaction itself (the means of payment). It is also the medium that allows one country to settle its obligations with, or transfer wealth to, another (world money). Finally, money is the medium for forming hoards, which are possessed by individuals or enterprises and held with banks or other financial institutions (means of hoarding). Financial institutions also hold their own vast hoards of money (reserves).

Money also has broader social functions in a capitalist society, most clearly seen in relation to power and hierarchy. Money affords social power, since it can impel others to comply with its owner's will, for example, by placating opponents, mobilising supporters, or hiring professional expertise. Money also affords political power, as is clearly seen in the influence exercised on political parties by those that finance them. Money, moreover, determines rank and social hierarchy, since it opens the doors of 'good' society and secures membership of exclusive clubs and associations. In capitalist society, which typically shuns hereditary distinctions and privileges, money is uniquely able to sustain rank and hierarchy across the generations, since it can place one's children in the 'right' schools and purchase husbands and wives.[1] Finally, money's power is also global since it allows countries to acquire military weapons produced by others, and since countries that make gifts of money can also persuade others to do their bidding.

The complex economic and social functions of money are matched by a bewildering array of its forms. There is gold, which lies mostly in private and public hoards. There are cheap metallic coins and banknotes used heavily in the petty transactions of everyday life. There are many different types of bank deposits that can be used to effect payments, or transfer wealth, among individual and large corporations. There are bank and other accounts that can be charged through the use of credit cards. There are also deposits held by financial institutions other than banks that can be used for payment. There are, moreover, several credit instruments that can be used in lieu of payment with cash, such as commercial bills. Despite money's protean aspect, however, the vast bulk of its forms in a developed capitalist economy have one thing in common: they are related to the credit system. The bulk of modern capitalist money is credit money.

The social relations captured by money in commodity exchange

These simple observations about capitalist markets and money appear unobjectionable, what economists call 'stylised facts'. Consequently, it comes as a surprise to find that mainstream economic theory leaves little room for money in its analysis of markets. To be sure, there are standard references to money's functions in economics textbooks, but they sit very uneasily with the underlying analytical approach of mainstream theory. The theoretical model of 'general equilibrium', which underpins mainstream economic thinking, is fundamentally a model of direct commodity exchange between market participants (Hahn 1982). Mainstream economic analysis, which prides itself in being the most advanced social science, at bottom sees capitalism as a social system in which things exchange directly for other things (barter), rather than for money. In short, mainstream economic theory analyses capitalist markets without adequately explaining money's role.[2]

Marxist political economy is vastly different on this score: money is shown to emerge spontaneously and necessarily whenever regular commodity exchange is undertaken. It is deeply misleading to assume, as mainstream economics does, that widespread commodity exchange could take place under barter conditions. There is no evidence (historical, anthropological or sociological) that a durable system of entirely money-free commodity transactions has ever existed. Indeed, research into exchange systems in which commodity owners regularly and frequently meet each other shows that money is present and touches all transactions, directly or indirectly.[3] Economic interactions between owners of particular commodities inevitably lead to the emergence of money as the universal commodity, the 'independent form of value' or 'universal equivalent'. Money and markets are inseparable.

In the first volume of *Capital*, Marx (1867, ch. 1) provided the building blocs for a theoretical explanation of money's emergence as part of his discussion of the 'form of value'. Money is shown to emerge spontaneously and inevitably whenever commodity owners come into frequent contact with each other. A very important point here is that money does not emerge simply as a generally accepted means of exchange, the mere lubricant of markets. Rather, money is the 'universal equivalent' or 'independent form of value'. Its essential property is that it can be immediately exchanged for all other commodities, thus enabling its owner to buy all others. Money emerges in commodity exchange as the monopolist of buying power; it is a special commodity that possesses a unique ability. The process through which this takes place is determined by the social relations between commodity owners, analysed below.

Markets are places in which independent and separate individuals interact with each other. Market participants might be related through kinship, friendship or social habits, but when they meet each other as commodity owners, these links recede into the background. They do not fully vanish, but become dominated by the characteristics of commercial give and take, by the 'bottom line'. The overwhelming concern of commodity owners when they meet is to obtain the exchange value of their commodities, to secure the 'quid pro quo' of value that is the very logic of their market activities. As far as this purpose is concerned, other market participants are strangers, alien individuals with whom a social relationship is to be constructed in the market alone. Thus, whenever two commodity owners meet (the 'accidental form of value'), one must make the opening move in establishing a social relation between them: there has to be an initial gambit. Typically, this takes the form of making an offer to sell the commodity possessed. The counter-party is, thus, given the option of accepting or rejecting the offer. The social relation that begins to emerge between the two commodity owners places the former in the position of the 'relative' or 'active' and the latter in the position of the 'equivalent' or 'passive'. To put it differently, when two alien commodity owners meet and begin to interact with each other, one of them immediately places the other in the position of being able to buy, even if only one commodity. Their social relation, defined as it is by the market, unfolds on this basis.[4]

Emergence of money represents the development of this rudimentary ability to buy, and its monopolisation by a single commodity. It occurs as transactions take place generally and frequently among similarly independent and separate market participants. As they meet each other and engage in quid pro quo transactions, their social relations develop further and revolve around a single pole of buying ability. There are successive steps to this process. First, one commodity owner makes an offer of sale to many others (the 'expanded form of value'), giving to all of them a little of the ability to buy (making them partial 'equivalents'). Second, in reverse, many commodity owners make offers of sale to a single other (the 'general form of value'), giving to the latter a much strengthened ability to buy (making the commodity involved a 'universal equivalent'). Third, if a commodity has come to possess exceptional buying ability (it already is the 'universal equivalent' for a group of commodities), still other commodity owners offer their commodities for sale against it because of its power to buy and not because they want to consume it. Its ability to buy increases correspondingly. On this basis, one commodity eventually attracts toward it offers of sale from all other commodity owners, becoming an 'equivalent' for all others. This is money, the commodity that can buy all others. It can do so simply because all other commodities are typically offered for sale against it.

When transactions become monetary, the social relations among commodity owners acquire a different content. Commodity owners are still independent and separate from each other, but they also act in a social way (if unplanned and unconscious), since they make money emerge by collectively offering their goods for it. Thus, money has its roots in individual exchange transactions among alien individuals, but it is also a collective and social phenomenon. Commodity owners typically offer their goods for money because they know that money will also be accepted by others. In short, money is systematically used by market participants because its use has become a social norm that characterises markets. However, the general use of money is a very peculiar social norm. It links essentially alien individuals and does not rest on familial, religious, hierarchical relations on which social norms typically depend. Money is the glue that holds together the individuals that comprise the market, it is the *nexus rerum* of a market economy. But it is an impersonal link, lacking the immediacy and directness of other norms that hold society together. Participants in capitalist markets are inherently separate from each other, their connections established by a thing that monopolises buying ability, the use of which has become a social norm.

The source of money's social power and influence is now clear. Contrary to what is typically (though often implicitly) assumed by mainstream economic theory, markets are not characterised by equality among participants. One commodity stands above all others, since it possesses the unique characteristic of being able to buy all others. Far from being democratic and egalitarian, markets have a privileged king and a vast crowd of subjects. This is the foundation of the power possessed by money owners compared to plain commodity owners. Money owners can mobilise resources, obtain commodities, secure promises and postpone demands on them in ways not available to plain commodity owners, thus affording to themselves economic power. In societies in which commodity exchange is widespread, the economic power afforded by money naturally leads to social power. In a capitalist society, which incorporates a vast network of markets, the king of the market is a prime instrument for imposing one's will on others, and establishing social hierarchy and rank. Social power, privilege and inclusion in various activities are intertwined with possession of money in a capitalist society. Equally, lack of money translates into powerlessness, deprivation and exclusion from several social activities for the majority of the poor in capitalism. In capitalist society, successful participation in social affairs depends less on a person's abilities and skills and more on possession of money.

Forms and functions of money

Thus, from Marx's work it is possible to piece together an explanation of what money is, namely the monopolist of the ability to buy in markets, 'the universal equivalent'. Money emerges necessarily when commodity owners interact with each other and, in turn, its use becomes a social norm. Does this derivation imply that money has to be a commodity? Is Marxist analysis of money tantamount to a theory of metallic money, namely gold, as Schumpeter, the great Austrian economist, thought (1954, pp. 699–701)? Moreover, since commodity money plays a marginal role in the contemporary world economy, is Marxist analysis obsolete? These questions sound plausible but actually reveal confusion regarding what money is, its corresponding functions and the forms it takes when it performs them.

The first point to stress is that the multiple economic and social functions of money flow from what the 'universal equivalent' is, namely the monopolist of the ability to buy. For the functions of measure of value and means of exchange to become real economic phenomena, the money owner has to accept an offer of sale from the commodity owner and part with money. A theorist can certainly create abstract models of value being measured and commodities being exchanged in a variety of ways, but for these functions to become social reality, there must be regular payment of money in exchange for commodities. In short, if the ability of money to buy is not exercised regularly in practice, the functions of measure of value and means of exchange have no social content at all. The same holds true for the hoarding and paying functions of money, domestically and internationally. It is possible for commodity owners to create obligations among themselves that rest on subsequent use of money as means of payment because money's monopolistic ability to buy makes later payment in practice acceptable. Similarly, commodity owners hoard money in order to be able to confront unforeseen events in the markets because money has a unique ability to buy. The multiple functions of money rest on its monopolisation of the ability to buy.

Money's original form has to be that of a commodity. It cannot be otherwise since money emerges within a set of commodities as the monopolist of buying ability. But as commodity money performs the function of means of exchange, symbolic money begins to emerge. By being used in exchange, commodity money is abraded and worn, and thus has less weight than it purports to do. Through use, metallic money spontaneously turns into a symbol of what it is supposed to be, and opens the way for proper symbols of money (paper or metallic) (Marx 1859, pp. 108–14). Furthermore, by performing the function of means of payment, money allows growth of trade credit ('buy now – pay later'). Similarly, the hoards created by money make it possible for their owners to make loans aimed at earning interest, thus opening the possibility of money-lending credit. In a capitalist economy, financial institutions emerge that make systematic the advance of both types of credit. Through their operations a proliferation of other forms of money takes place, all of which are essentially credit money.

In all its forms (commodity, symbolic or credit) money remains the 'universal equivalent', the monopolist of the ability to buy. At the same time, it cannot be assumed that every new form of money is fully adequate for the particular function that it is called to perform in exchange. Monetary problems and crises may occur if the form is inadequate for the function, for example, price inflation may arise out of fiat and credit money functioning as means of exchange. In developed capitalism, the functioning of commodity money has been limited to hoard of last resort. Such hoards are held by the major financial institutions (central banks) of the capitalist credit system. The marginal role played by commodity money in advanced capitalist exchange poses no insuperable problems for Marxist political economy.[5]

MONEY AS CAPITAL

Money and the circuit of capital

Money is not a specifically capitalist economic phenomenon. The presence of money and its extensive social and economic functioning are well attested in ancient societies as well as in contemporary communities that are in no way capitalist. Given the analysis of money as monopolist of the ability to buy, it follows that money's presence and functioning in non-capitalist societies depends on the extent to which commodity exchange is present in them. Nevertheless, money's nature, functions and forms emerge most clearly under capitalist social conditions, for it is only then that commodity exchange becomes truly general and permeates economic activity.

There are two reasons why commodity exchange and money occupy such a prominent position in capitalism compared to other societies. First, capitalist production is undertaken by a class of autonomous and competing producers (capitalists), who purchase inputs and sell output in a range of markets. Capitalist production, moreover, relies on the social class of wage workers. They derive their income from selling their ability to work in the labour market, and use the proceeds to obtain means of consumption in commodity markets. The prominence of markets in the economic functioning of capitalism ensures the prominence of money's economic and social role. Second, as discussed elsewhere in this volume, the existence of a capitalist class and a working class turns commodity value into a deeply rooted social norm. The economic interaction of these two classes gives to value a real social substance, namely abstract labour. The driving motive and mainstay of capitalism is the continuous expansion of value as abstract labour, through extraction of surplus value from workers employed in production. Since it is the independent representative of value, money possesses a special role in capitalism: it captures its very social essence, summarised in the drive for money profits.

The special place occupied by money in capitalism is shown by money becoming capital. Money as capital is a broader social and economic phenomenon than money as money. For Marxist political economy, capital is the sum total of social relations between capitalists and workers, but also the ceaseless movement of value in pursuit of self-expansion. The latter is best thought of as a circular flow: capital value starts as money, becomes material inputs for production through market purchases (means of production and labour power), turns into finished commodities through production, and returns to money (augmented by surplus value generated in production, i.e. profit) through sale of finished commodities (Fine 1975). Money is the natural starting and finishing point of this circuit. Since it is the circuit's most fluid element, money is the form in which capital normally commences its movement as capitalists make investment purchases. It is also the form to which capital must return (plus profit), if the capitalist is to retain the ability to invest where profit can be maximised. Money as capital can be indicated by $M - C - M'$ (money – commodities – more money), a summation of the circuit of capital. This is in contrast with money as plain money, indicated by $C - M - C'$ (commodities – money – other commodities), a summation of market transactions (or simple exchange). The motivating purpose of $M - C - M'$ is acquisition of money profit, while that of $C - M - C'$ is acquisition of different use values. Money provides the objective of $M - C - M'$, but also a means for achieving this objective since money as capital hires workers and allows for generation of surplus value (see Chapter 1). Yet, money as capital takes advantage of – and does not eliminate – money's functioning as plain money. The point is important for reasons of both theory and policy, as can be seen in the following two ways.

First, the profit-seeking and exploitative character of capitalism revolves around money but does not result from money's peculiar properties. Rather, capitalism originates in the profound social transformation that creates the social classes of capitalists and workers. It is true that for the emergence of capitalism extensive use must be made of money's capacity to be hoarded and to pay in order to create the original capital available to the capitalist class as well as dispossess the working class from the means of production. But the driving force behind these changes is social struggle, of which money is a means and not a cause. Put differently, it is not money that creates capitalism, but capitalism that transforms money into capital. Capitalist social relations graft onto money's functions as means of purchase, payment and hoarding the aspect of capital, especially since surplus value takes the form of money profits. These functions are incorporated into the circuit of capital and facilitate the ceaseless expansion of capital. It follows immediately that to tackle the roots of capitalism and stop money from functioning as capital it is not enough to confront the monetary mechanisms of capitalism. Instead, it is vital to challenge the class relations that underpin it and result in exploitation. To stop money acting as capital it is necessary to change the class structure of capitalist society rather than simply disrupt its monetary mechanisms.

Second, in developed capitalism the economic and social space of simple commodity exchange (C – M – C') expands dramatically. From the standpoint of the working class, market transactions are simple circulation: workers enter the labour market, sell their ability to labour, and use the money to obtain necessary means of consumption. Since the development of capitalism implies the expansion of the working class (that is, the class that earns income by entering the labour market), it follows that as capitalism develops, the functioning of money as plain money is intensified. Money as plain money characterises transactions relating to the sale of labour power, and that is how it enters the realm (and consciousness) of the worker. The driving motive for workers in such transactions is acquisition of the use value of goods, and has nothing to do with the expansion of value. For workers, money is primarily the means of purchase and of settling obligations, and in a very limited sense, the means of hoarding.

Consequently, money as the monopolist of the ability to buy directly affects the social power of workers in capitalism – and of the poor generally. Their social power could increase dramatically if the buying power of money was limited relative to key goods. This is not conditional on changing the class structure of capitalism, or on overthrowing it. It simply follows from limiting the monopoly power of money over important elements of the consumption of workers and the poor. When access within capitalism to health, education, and transport is regulated through public provision rather than through private expenditure of money, the social power of workers rises sharply. Public provision of such goods and services is not only more economical but also makes for greater social power and confidence for those who have limited access to money. Money as capital has little to do with this result. What is vital is to restrict the functioning of money as money in the social realm of workers and the poor.

Money and the credit system

Money's role in the circuit of capital brings one more fundamental change to its social and economic functioning: money is systematically mobilised in credit and finance. Credit practices, that is, both the advance of goods on trust (for later settlement of the obligation) and the lending of money, are found in a wide variety of non-capitalist societies. However, in those societies the practices of credit are peripheral to the main activities of production, being aimed mostly at facilitating or smoothing consumption. No mechanisms for the systematic lending of money to undertake productive investment can be found in non-capitalist societies (Itoh and Lapavitsas 1999, ch. 3). In contrast, capitalism contains a financial system, a vast and elaborate social structure that puts credit and finance at the service of capitalist production.

Money's role in the circuit of capital is of critical importance for the capitalist financial system in two related ways. First, by functioning as means of payment, money allows for the systematic advance of finished commodity output against promises to pay. Thus, money makes possible the expansion and growth of trade credit among capitalist enterprises. The typical way of undertaking market operations in developed capitalism is on trade credit rather than cash, because such credit economises on money capital and speeds the turnover of capital. Second, by functioning as means of hoarding, money allows for systematic concentration of idle money in the course of the circuit, and creation of loanable money capital. Hoards are systematically formed by capitalist enterprises as precautionary reserves, fixed capital depreciation, reserves necessary for maintaining the continuity of production, and so on.[6] Hoards are also formed as workers and capitalists realise their consumption through money. The financial system gathers money hoards across society and turns them into loanable money capital. This is a special form of capital, which does not earn profit through direct engagement in production and circulation but earns interest by being lent. Access to loanable money capital allows capitalists to start new – or to expand existing – circuits of capital, thus increasing the mass of surplus value generated by their own capital. Interest is a share of the additional surplus value, which accrues to the owners of loanable money capital (see Chapters 2 and 4).

The capitalist financial system is set complex social mechanisms that organise trade credit, mobilise loanable money capital and transfer money across society. Its operations rely on money. The hoarding function of money allows reserves to be created that can form loanable money capital, by definition impossible in the absence of money. The paying function of money, on the other hand, allows for systematic cancelling out and residual settlement of obligations among capitalists (clearing), which encourages growth of all credit practices. The paying function is also vital to lending, as money can reliably transfer value to claimants at specified points in time, whether as interest or principal. In turn, the form of money is profoundly affected by growth and development of the capitalist financial system. Banks and other financial institutions systematically generate credit money that overtakes commodity and state fiat money as 'universal equivalent'. Capitalist money is overwhelmingly credit money, mostly functioning as means of hoarding and payment. The means of exchange function is relegated to the small change of credit money (mostly banknotes).

The financial system represents a concentration and expansion of social power on quite a different level from mere money. Access to credit enables capitals to move into different areas of production and beat others in competition. The financial system distributes spare resources across society, hence control over its mechanisms matters greatly for the direction of development of a particular society. Nevertheless, the enormous social power that is afforded by the financial system cannot be analysed in the context of money – it requires discussion of the social relations of credit and finance which, despite having a monetary aspect, are very different from the social relations encapsulated in money (see Chapter 10).[7] One point that should be made in this connection, however, is that the deeper foundation of the financial system in capitalism can be found in the systematic generation of surplus value in the circuit of industrial capital. Surplus value allows for systematic payment of interest and provides the wherewithal for other returns made by capitals engaged in finance. Capitalist society is the only historical society that has been able to evolve a financial system, as opposed to simple credit transactions, because it is the only society that systematically generates money profits in production. Thus, although the power of finance is enormous in a capitalist society, ultimately finance is subservient to industrial capital.

CONCLUSION

Money is an economic category intrinsic to markets and fundamental to relations between commodity owners. It arises spontaneously in commodity exchange, through the social (but unplanned and unconscious) action of other commodity owners. It is the monopolist of the ability to buy, or in Marxist terminology, the 'universal equivalent'. Money has several complex functions vital to commodity exchange – measure of value, means of exchange, means of hoarding, means of payment and world money. As it performs these functions, money's own form is altered into forms that include commodity money, symbolic money and credit money. Each form of money has to be adequate for the function that it tends to perform. Money's unique ability to buy gives it an exceptional position in commodity markets. Hence, access to money becomes a source of economic and social power, and the foundation of capitalist hierarchies and privileges. It also follows that the social power of working people and the poor in capitalism would benefit from limiting money's ability to buy, especially over the goods that significantly affect their living conditions.

Money is an economic category that is far older than capitalism. Nevertheless, its nature and functions emerge most clearly under capitalist social conditions because it is then that commodity exchange becomes truly general. Moreover, under capitalist conditions, money becomes capital. It is both starting (money investment) and finishing point (sales revenue) of capital's characteristic circular movement. More importantly, money provides the motive (money profit) for capital's operations, and captures its essential purpose: self-expansion. Since it can be used to hire workers necessary for generation of surplus value, money is also the means through which capital can bring about its self-expansion. Consequently, the social power of money in capitalism is enormous. Under capitalist conditions, furthermore, money becomes one of the foundations of the financial system by allowing trade credit to proliferate and by making possible the formation of loanable money capital. Its close association with the financial system induces broad changes in the form of money, and credit money becomes the characteristic form of money in capitalism. The power of the financial system over capitalist society is also enormous, but the social relations of finance need a broader framework of analysis than those of money.

REFERENCES AND FURTHER READING

Fine, B. (1975) *Marx's Capital*. London: Macmillan.
Fine, B. and Lapavitsas, C. (2000) 'Markets and Money in Social Science: What Role for Economics?', *Economy and Society* 29 (3), pp. 357–82.
Hahn, F. (1982) *Money and Inflation*. Oxford: Blackwell.
Itoh, M. and Lapavitsas, C. (1999) *Political Economy of Money and Finance*. Macmillan: London.

Kiyotaki, N. and Wright, R. (1989) 'On Money as a Medium of Exchange', *Journal of Political Economy* 97, pp. 927–54.

Lapavitsas, C. (2000a) 'Money and the Analysis of Capitalism: The Significance of Commodity Money', *Review of Radical Political Economics* 32 (4), pp. 631–56.

Lapavitsas, C. (2000b) 'On Marx's Analysis of Money Hoarding in the Turnover of Capital', *Review of Political Economy* 12 (2), pp. 219–35.

Marx, K. (1859) *Contribution to the Critique of Political Economy*. Moscow: Progress Publishers, 1970.

Marx, K. (1867) *Capital*, vol. 1. London: Penguin, 1976.

Menger, K. (1892) 'On the Origin of Money', *Economic Journal* 2, pp. 239–55.

Schumpeter, J.A. (1954) *History of Economic Analysis*. London: Routledge.

Uno, K. (1980) *Principles of Political Economy*. Brighton: Harvester.

NOTES

1. For a fuller discussion of money's social role and power see Fine and Lapavitsas (2000).
2. There have been neoclassical attempts to explain the spontaneous emergence of means of exchange as the most 'marketable' commodity, going as far back as Menger (1892). The most recent formulations of this idea, for instance, Kiyotaki and Wright (1989), leave the property of 'marketability' unexplained. In effect, money is the most 'marketable' commodity because market participants think that it is. That is a deeply unsatisfactory and circular argument.
3. See Itoh and Lapavitsas (1999, chs. 2 and 10).
4. Positing Marx's analysis of money in these terms is one of the most decisive contributions of the Japanese Uno school (Uno 1980). It is not implied here that Marx's analysis of money is the final word on the subject. The point is, rather, that it offers a path toward solving the 'riddle of money', while also taking into account the social relations encapsulated in money.
5. For further analysis of this issue see Lapavitsas (2000a).
6. Hoarding in the circuit of capital is fully discussed in Lapavitsas (2000b).
7. See Itoh and Lapavitsas (1999, chs. 3 and 4).

4 Capitalist Competition and the Distribution of Profits

Diego Guerrero

Universal competition among all those who sell commodities and depend on their sale, as well as the capitalist distribution of the output, must be understood and analysed together in the framework of their own mode of production: capitalism. The mode of distribution of the social product is a consequence of the actual mode of production. When capitalism prevails, its main feature is the all-embracing dependence of the social processes (including the labour process) on the specific way production is undertaken. Capitalist production is carried out in a private and socially fragmented way, with no possibility of systematic co-operation beyond each unit of production (see Chapter 1).

THE ALL-EMBRACING COMPETITIVE STRUGGLE AND THE PRIMARY DISTRIBUTION OF INCOME

The fragmentation of social production into private, independent and rival units reaches its maximum when labour power becomes a commodity. Then the wage workers and the public administration, as well as the capitalists, behave as merchants. The workers depend on the sale of their labour power, and the state – whose revenues derive from a productive sector that produces commodities as the sole means of making money – follows a similar merchant behaviour. Hence, wherever rivalry and competition form the system's status quo, all agents (workers, capitalists, and the state) must behave as merchants subject to the rules of the competitive war. The study of these rules is the core of the theory of competition, an aspect of value theory bearing upon the distribution of the means of production, its implications for the primary distribution of the newly produced value (between variable capital and surplus value) and, especially, the distribution of surplus value among its co-sharers (see below).

The class struggle itself, although not reducible to competition, includes a competitive dimension. However, the reproduction costs of simple labour power are the main determinant of its normal price (the wage rate). The necessity that this reproduction should be accomplished without jeopardising the continuity of the process of capital accumulation ensures that 'subsistence wages' (in a social rather than physical sense) remain the norm in contemporary capitalist economies (see Chapter 5). This 'subsistence' level includes all categories of wage labour. The fact that the flow of former capitalists (and self-employed workers) becoming new wage workers is greater than the opposite flow is explained by the fact that the threshold (money) capital required to set up a new capitalist firm is growing faster than the monetary reproduction costs of the average socially qualified worker. The net result of this process is the growing proportion of wage (or proletarian) labour in capitalist societies (see Table 4.1).

Table 4.1   Proportion of Waged (Proletarian) Labour Power, in Selected Countries and Years.

Country	1930–40	1974	1997
USA	78.2 (1939)	91.5	91.5
Japan	41.0 (1936)	72.6	80.8
Germany	69.7 (1939)	84.5 (West)	90.7
UK	88.1 (1931)	92.3	87.3
France	57.2 (1936)	81.3	87.6
Italy	51.6 (1936)	72.6	74.7
Canada	66.7 (1941)	89.2	n/a
Belgium	65.2 (1930)	84.5	83.6
Sweden	70.1 (1940)	91.0	94.7
Spain	52.0 (1954)	68.4	81.0
Europe – 15	n/a	n/a	84.3
Simple Average	65.2	83.7	86.2

The trend towards a growing relative immiseration of the workers is well documented, and it should not be confused with the simultaneous trend toward an increasing real wage. The two trends are not only mutually compatible but each is inherent in capitalism, as can easily be seen in developed capitalist societies. As was pointed out by Marx (1867), the increase in labour productivity reduces the labour value of each commodity, and each bundle of commodities (including the 'subsistence' bundle of the workers). At the same time, the increase in average labour intensity generates a trend towards higher consumption levels, as the only way of replenishing the increased labour power consumed per hour. In fact, this duality – rising real wages and declining 'relative' wages – is a very important factor conditioning the behaviour of the working class. The workers can improve their material standard of life in the long run (even if going through phases in which their purchasing power stagnates or even declines) while, at the same time, inequality grows in terms of the relative position occupied by the working class in contrast to its antagonistic class: the capitalists.

Edward Wolff (1998) has shown that the net financial wealth of the average family in the United States is ten times *smaller* if 'only' 99 per cent of the population is taken into account (leaving aside the highest 1 per cent). It can also be shown that, in several OECD countries, the rate of surplus value (the rate of exploitation) has been rising for two centuries. The only categories we need are those of surplus value, exploitation and others derived from the labour theory of value (for a review of the literature, see Shaikh and Tonak 1994).

COMPETITION AND PROFIT DISTRIBUTION BETWEEN FIRMS IN THE PRODUCTIVE SECTOR

Each capitalist firm gathers a mass of workers into a single operating mechanism called its 'collective' labour force. In this system of production, work is collectivised at the level of the individual firm, but it cannot be co-ordinated with the remaining social labour in the framework of the capitalist mode of production. 'Direct' labour performed by the whole 'collective worker' (the sum total of the collective workers in all firms) produces an amount of new value greater than that needed to reproduce the value of the collective labour power (the value of their means of subsistence or regular consumption). This is due to the generalised existence of surplus labour, that is, labour over and above the amount needed to reproduce the equivalent of the bundle of goods actually consumed by the direct producers. The monetary expression of the surplus labour appropriated by the owners of the firms is the total surplus value, or profit, extracted by the capitalist class. The core of the theory of competition concerns the allocation of this surplus value and, specifically, the discrepancies between the 'individual' amounts of surplus labour extracted and realised by each of the rival productive units.

The state and other institutions able to modify the basic results of the free competition model should be temporarily set aside, so that we can focus on productive capital only. Accordingly, we exclude taxes and monopolies, and the existence of goods that are not freely reproducible by manufacturing (land, for instance; see below). We will firstly study competition among capitalist firms as a process which is conveniently split into two different analytical moments: *intrasectoral* competition and *intersectoral* competition. Even if both take place simultaneously in practice, they should be analysed separately and successively in order to facilitate understanding (see Chapter 1 and Gouverneur 1983).

Intrasectoral competition

Intrasectoral competition occurs between firms belonging to one sector, i.e. all those producing the same kind of commodity (homogeneous product). Technical diversity within each sector makes the unit production costs very different in each of the firms. However, all of them are forced to accept the tendency toward the same output price, and not demand a higher one, due to their competition for market shares. These different unit costs, and the simultaneous tendency toward homogeneous prices, generate a tendency toward the dispersion of the individual profit rates obtained by each firm.

However, it is crucial not to confuse the *cost per unit of input used up* with the *cost per unit of output produced*. This is a very important issue, as intrasectoral competition frequently takes place in a worldwide framework, and the firms producing the same type of commodity face an increasingly globalised market. The competitiveness of a firm, like that of a sector or country, is ultimately based on an advantage in unit costs. If the price of a unit of labour power employed in sector S is lower in country A than in country B (say one half), but labour productivity is much higher in B (say six times higher), the result will be that the wage cost per unit of product will be three times lower in country B (even if the wage rate is higher in this country). If both producers face approximately the same input prices, their profit rates will be very different and, paradoxically, they will be higher in the high-wage country (since high wages usually reflect productivity differences).

Intersectoral competition

Intersectoral competition operates between firms belonging to different branches or sectors. As Marx (1894) points out, when taking into account the fact that commodities circulate not simply as commodities, but as the product of capitals (i.e., as capitalist commodities; see Rubin 1928), competition requires that any amount of capital invested in one sector should gain a proportionate yield (an equal profit rate). This means a profit that tends towards proportionality to the sum of its variable and constant capital, in spite of the composition of capital in each sector (the ratio between the two components of capital) being very different in each of them.

Both dimensions of competition produce quantitative modifications in the value of the individual commodities and in the profits received by individual capitalists. The latter happens even if the total value and surplus value produced are unaffected by this double redistribution. Marx (1894) insisted that the unit value in each sector can be modified as a result of 'free competition among capitals'. Free competition (as was pointed out by Smith 1776) prevents one sector from obtaining a higher average profit rate than the economy's average, since the search for maximum profit rates by each individual capital generates a tendency toward the equalisation of the average rate of profit in every sector. Marx explained that these 'modified' or 'transformed' prices, arising from this second tendency in competition – what he called 'prices of production' – would not be strictly proportional to the total amount of labour spent in production. This is because differences in the organic and value compositions of capital between sectors require that, in the context of intersectoral competition, profit should be proportional to the total capital invested, rather than proportional to its variable component only (the fraction of capital exchanged against the only commodity capable of producing surplus value, labour power).

Before proceeding to the next section, it is necessary to add two considerations. First, even if Marx considered Smith's treatment of the tendency toward the equalisation of the sectoral rates of profit to be one of his most important contributions, he completely rejected the ideological (normative) conclusions that the apologists of capitalism extract from the idea of the 'invisible hand'. Marx distinguishes between two things. On the one hand, it is true that supply tends to adjust itself (more or less slowly) to the demand existing in actual capitalist conditions: this is the 'automatic' mechanism in capitalist reproduction, allowing the pursuit of individual interest on the part of each firm to lead to a certain mode of social reproduction. However, there is no guarantee that this effective demand truly reflects the needs of the members of society, for it is simply a monetary demand expressing the mode of distribution corresponding to a system of production that reproduces wealth *and* poverty in both poles of the same basic (capitalist) relationship. Moreover, although the prices of production are the centres of gravity regulating the movement of actual (market) prices in conditions of 'free competition', the existence of monopolies or public intervention, acting through 'price regulation', may alter the normal oscillation of these prices around their regulating centres, determined by the conditions of 'free competition' (this process should not be confused with the neoclassical theory of perfect competition).

THE DISTRIBUTION OF PROFIT OUTSIDE THE PRODUCTIVE SECTOR

The presentation of the theory of competition based on the labour theory of value is not yet complete. We should now deal with the unproductive sector of the economy, especially the state (which finances itself through taxes and other revenues originating from the productive sector) and the circulation activities (as opposed to the productive sector), including the redistribution of part of the surplus value (or its money form, profit), and land rent or, more generally, any kind of payment made for the use of inputs that are not freely reproducible.

The public sector

The state – leaving aside the public utilities, which should be dealt with exactly like private firms that, in this regard, belong either to the productive or the circulation sectors – supplies the so-called 'public services' usually without any merchant transaction or price. This means that the state must take up a fraction of the profits generated in the productive sector of the economy, in order to pay for the expenses generated by its 'administrative' activities (both when it performs the most useful activities, like public health or education, and when it shows more clearly its capitalist class nature, as in the defence of private property or in helping to fund private firms). The taxes levied by the state and other public institutions (in a broad sense, including fees, social security contributions and other revenues) are a fraction of the total surplus value that cannot be directed toward the ultimate aim of the capitalist class: accumulation, as additional capital available to expand the scale of operation of the productive sphere. Hence, it must be considered as a form of 'social consumption' of part of the output of the productive sector.

Commodity circulation

The state is not the only sphere where unproductive labour is performed (i.e., labour creating neither value, nor surplus value, nor capital; see Chapter 2). The 'sphere of circulation' must be clearly distinguished from the sphere of production, since only the latter creates the whole mass of value, while the sphere of circulation displaces and distributes that mass without modifying it. This difference is crucial, since the analysis of exploitation starting from the labour theory of value ought to be based on the assumption of equivalent exchange. What this means is that, in the global process of capitalist production, $M - C \dots P \dots C' - M'$, value and surplus value (hence profit) are created only in $\dots P \dots$, the phase of production, whereas in both circulation processes (the purchase of inputs, $M - C$, and the sale of the output, $C' - M'$) all that happens is the transfer of ownership from the seller to the buyer, without any modification in the value of the commodity exchanged (see Chapter 1).

Empirical works dealing with this issue can be misleading, in incautiously identifying the concept of 'circulation' with what the available data characterise as the 'trade' and 'finance' sectors. In my view, Nagels (1974) has correctly insisted that this should be avoided and, instead, that researchers should attempt to analyse two problems that are often ignored: (a) that productive activities are performed in these sectors (see Guerrero 1999–2000), and (b) that it is necessary to locate unproductive circulation activities inside all productive sectors, because capitalist economies produce not only goods but *commodities* and, hence, needs to transmit titles of ownership, as well as perform other activities that are superfluous, from the point of view of the use value produced and its consumption.

Land and other non-reproducible inputs

Finally, the question of land rent requires a special treatment in the theory of value, competition and distribution (Bina 1985). Productive inputs privately appropriated and reproducible only in a limited way allow their owners to participate in the distribution of the surplus value created by the workers in the productive sector. The reason is simple: these owners can claim from the productive capitalists a share of the total surplus labour, and this share increases with the demand for these inputs (whose supply is, necessarily, limited). Marx (1894) wrote that 'the fact that capitalist ground-rent appears as the price or value of land, so that land, therefore, is bought and sold like any other commodity, serves some apologists as a justification for landed property since the buyer pays an equivalent for it, the same as other commodities ... The same reason in that case would also serve to justify slavery, since the returns from the labour of the slave, whom the slave-holder has bought, merely represents the interest of the capital invested in this purchase' (p. 642).

Marx criticised Ricardo (1821) for analysing differential rent only. Instead, for Marx, there is also an 'absolute rent' alongside the former. Absolute rent is appropriated by the landowners whenever the demand for the commodity produced with help from land (or other non-reproducible inputs) raises its price above zero. Absolute rent is simply due to the 'monopoly of the ownership of the land', and this 'limitation' to the free circulation of capital (and hence to the general theory of competition) 'continues to exist even when rent in the form of differential rent disappears' (Marx 1894, p. 751). In contrast, differential rent benefits the owners of land (and other limited resources) who are in a better position relative to their fellow landowners, either due to the better quality of their land (rich soils for agriculture, better weather in land for tourist uses), proximity to the place of manufacturing or sale of the output, or easier exploitation (in the case of mining or exploiting underground or marine deposits or urban land) and so on. In this way, the owners of the best quality non-reproducible resources make possible production at a lower cost than that included in the normal (production) price, and appropriate the difference.

What has been said in the previous paragraph applies to the so-called 'differential rent I'. Marx also discusses 'differential rent II', which arises as a consequence of an additional investment of capital on a given plot of land, keeping constant both the productivity differential of this allotment with respect to others, and the regulating price of the commodity which is being produced with the help of this land.

Consequently, in the case of land and other non-reproducible resources, it is the conditions of the *least* efficient units that regulate the price of the commodities to which these inputs contribute. This is the opposite of what happens with the regulating capitals in most industrial sectors. In mature sectors, the regulating capitals are usually those enjoying the average conditions of production; in contrast, in sectors endowed with the most advanced technology, especially those undergoing rapid evolution (or 'revolution', such as the personal computer industry during the 1980s and 1990s), it is the most efficient productive units that set the normal price regulating the actual (market) price.

REFERENCES AND FURTHER READING

Bina, C. (1985) *The Economics of the Oil Crisis*. London: Merlin Press.

Gouverneur, J. (1983) *Contemporary Capitalism and Marxist Economics*. Oxford: Martin Robertson (updated French edition: *Decouvrir l'économie. Phénomènes visibles et réalités cachées*. Paris: Editions Sociales, 1998 see also the Spanish edition in <www.ibdoc.com>).

Guerrero, D. (1999–2000) 'Nonproductive Labor, Growth and the Expansion of the Tertiary Sector (Thirty Years after the Publication of *Marx and Keynes*)', *International Journal of Political Economy* 29 (4), pp. 14–55.

Marx, K. (1867, 1894) *Capital*, vols. 1 and 3. New York: International Publishers, 1967.

Nagels, J. (1974) *Travail collectif et travail productif dans l'évolution de la pensée marxiste*, Brussels: Editions de l'Université de Bruxelles.

Ricardo, D. (1821) *On the Principles of Political Economy and Taxation*, (3rd edn.), in *The Works and Correspondence of David Ricardo*, ed. P. Sraffa and M. Dobb, vol. 1. Cambridge: Cambridge University Press.

Rubin, I.I. (1928) *Essays on Marx's Theory of Value*. Detroit: Black and Red, 1972.

Shaikh, A. and Tonak, E. (1994) *Measuring the Wealth of Nations: The Political Economy of National Accounts*. Cambridge: Cambridge University Press.

Smith, A. (1776) *An Inquiry into the Nature and Causes of the Wealth of Nations*, ed. R.H. Campbell, A.S. Skinner and W.B. Todd. Oxford: Oxford University Press, 1976.

Wolff, E.N. (1998) 'Recent Trends in the Size Distribution of Household Wealth', *Journal of Economic Perspectives* 12 (3), pp. 131–50.

5   Contesting Labour Markets

Ben Fine

Standing alongside its undoubted achievements, there are some endemic features of capitalism that have persisted despite general, if not universal, condemnation and concerted attempts to mitigate and eliminate them. These include uneven development; poverty for many, even the majority, alongside huge wealth for a minority both within and between countries; deepening environmental damage; oppression by race, gender and ethnicity; and the apparent inevitability of armed conflicts. If capitalism has triumphed, much of its victory is hollow. At a deeper analytical level, such stark empirical realities concerning the contemporary world, as part of its continuing history, point to the systemic character of capitalism, and the presence of forces, structures, relations and processes that are not amenable to control. In a previous age, literally, a *deus ex machina* would have been invoked both to explain and to justify the complexities and contradictions of the real world, with mortals merely playing out a battle between vice and virtue according to a game set by divine rule. Now, in the age of reason, we cannot afford such ideological luxuries. They must be replaced by analysis.

At a less dramatic level, unemployment has shown itself to be uniquely characteristic of capitalism. Unlike other markets, even when the labour market is 'tight', it still leaves workers without jobs, leading mainstream economics to appeal to a 'natural' rate of unemployment in equilibrium, necessary for the economy to function smoothly. To some, this does not set the labour market apart from other markets, for all unemployment is perceived to be 'voluntary'. If only workers would offer themselves at a sufficiently low wage, they could be employed. They must prefer the leisure and other benefits attached to their chosen state of idleness. There are, of course, many objections to this view of the world, varying from the false picture painted of the unemployed themselves, often desperate for work, through to the various versions of Keynesianism that emphasise deficient aggregate effective demand as the cause of (involuntary) unemployment. As is readily recognised by those who care to see, recessions and unemployment do not reflect conscious choices, freely undertaken, but unconscious forces beyond our ken.

It is important, however, to recognise that those (Keynesian) critics who accept that unemployment can be involuntary share some questionable assumptions with their opponents. First is the idea that work necessarily incorporates what is termed disutility, and that it is a matter of the worker gaining maximum reward for minimum time and effort. Even within capitalist society, this is far from the full story since the waged worker's motivation is both complex and mixed. In addition, non-waged work, in the household for example or for recreation, is often undertaken for pleasure. In effect, one undoubted feature of *capitalist* employment – its often arduous and unrewarding nature – is taken for granted as an exclusive characteristic of *all* work.[1] Significantly, in his early work on alienation, Marx placed considerable emphasis on the uniquely dissatisfying nature of work under capitalism. He focused on the worker's loss of control over the production process, in conception, organisation and execution. Even if this is not the whole picture, the worker tends to become a repetitively rotating cog in a machine, as brilliantly displayed in Charlie Chaplin's film, *Modern Times*. And the worker has no control over the fruits of labour, the products themselves, as they belong to the capitalist. It is hardly surprising that other aspects of workers' alienation should be heavily contested under capitalism, in disputes over conditions of work and not just levels of wages. In general, workers seek more satisfaction from their work, and not just more pay for less time, but achievement of their goals is limited by the capitalist pursuit of profitability.

THE DISTINCTION OF LABOUR

There is then within mainstream economics a tendency to treat all work as if it were synonymous with work under capitalism (which is itself falsely conceived in terms of a simple trade-off between higher productivity for capitalists and lower disutility for workers). Hence the same theory is applied seamlessly across other forms of 'work', as in the new household economics and the economics of crime – focusing on the 'wages' of theft as against the disutility (derived from potential punishments). This is indicative of a more general drawback of economic theories of the labour market – they are universal, ahistorical and asocial. This is already apparent in the categories of analysis used – such as (dis)utility, production function, and labour itself. Whilst the theory is intended to address a labour *market*, it does so by deploying concepts that have no roots in such specific commercial circumstances. To return to the previous issue, of the inescapable presence of unemployment, it already presumes much that is to be examined and explained. For there to be the *un*employed, it is necessary to acknowledge that capitalist employment is the predominant form taken by work or labour, that a wage system is involved. In other words, we need to know what is different about the labour market in historical and social terms as well as by comparison with other commodities that do not experience chronic unemployment (a term that is used with extreme reluctance when describing markets other than labour).

Not surprisingly, there is a host of literature concerned with what is different about labour markets. It has spawned the disciplines and practices of industrial relations, human resources, and personnel management. These tend to focus on what is different about labour or what is different about the market in which it is bought and sold. As such, it does not deal directly with why labour takes the wage form and the significance of this in comparison with other markets. Economists have been even more negligent of such fundamentals, simply distinguishing labour by its conditions of supply and demand, like any other market. Significantly, the Nobel prize winner Solow (1990) deems it necessary to devote a book to persuading his fellow economists that the labour market is different from that for fish. This is a remarkable task to have set himself, not so much in its substance, but that it should be considered to be necessary. For economists have been reluctant to accept that labour markets are distinct from other markets. Essentially Solow's answer is that workers, unlike fish, represent themselves in the labour market. They have thoughts and feelings about fairness and fellow workers, for example, and can display these in terms of loyalty or resistance to a particular employer. Whilst humans and fish are different in these respects, a moment's reflection reveals that Solow has not otherwise distinguished between the two as *markets*. Fish, like all products, offer 'resistance' of one sort or another in being brought to the market, what mainstream economics would perceive to be the costs or conditions of supply. Moreover, fish *are* represented in the market by human agency, by those who sell the fish. Fishermen, fishmongers and others are also able to display and act upon motives of fairness and loyalty, in relation to one another as well as to those who stand on the other side of their market. Indeed, in more advanced (and often in the most primitive) labour markets, the worker is also represented by another, a separate agency, for example, by a trade union (or via family, kin or ethnicity).

More recently, a different approach to the specificity of labour markets, one that straddles mainstream economics and radical political economy, is the idea of efficiency wages. In its mainstream version, employers may choose to pay higher wages than necessary in order to secure a loyal, skilled and disciplined workforce. Lowering wages, even where there is unemployment, does not necessarily increase profitability because turnover, skills and work-intensity all suffer (the latter because the threat of the sack is lessened at a lower wage). This situation arises because of informational uncertainties – individual workers know how loyal, skilled and disciplined they are but bosses do not and may be willing to pay a premium on wages to obtain higher levels of these features on average.[2]

The radical version of efficiency wages differs little in analytical content from the mainstream version.[3] It does, however, add a richer interpretation in which the market inefficiencies arising out of informationally uncertain contracting are perceived to derive from the asymmetry between the two sides facing one another across the labour market. Conflictual contracting predominates over co-operative because of the capitalist ownership of the means of production. It is suggested that only if there were more equal and/or collective forms of ownership, then it would be possible for contracting to be more efficient and less antagonistic and for co-operation to prevail over conflict.

Irrespective of the latter's merits as a recipe for socialism (as a fairer contractual society), the distinctive nature of labour markets has yet to be captured by the efficiency-wage approaches. For informational asymmetries (the buyer not knowing as well as the seller what is being bought and sold) can be used to interpret *any* market and have, as such, been at the leading edge of mainstream microeconomics (often parading as macroeconomics). Indeed, the same analytical tools have been used in a whole range of applications, from the new financial economics through to the new development economics. Significantly, this way of looking at imperfect markets began with the example of 'lemons', slang for second-hand cars. This indicates that uncertain, conflictual contracting with asymmetric information and power does not get to grips with the nature of labour as distinct from other markets. As the old saying goes, 'Buyer beware!' What you get for your money has some uncertainty attached in whatever market you are buying, and not just in the labour market (and there is no guarantee of getting more or better by paying more although this is required by the efficiency-wage approach).

A more long-standing and common explanation for unemployment amongst political economists originated with Kalecki (1943) in what, in part, can be interpreted as the macro-version of the efficiency wage argument. Anticipating the potential impact of Keynesian policies in the coming postwar period, he argued that full employment could not be pursued. If it were, worker discipline would be totally eroded since threat of the sack would be empty given the ready availability of other jobs. There are, however, two problems with this account. First, it presumes that, should they want to, governments of capitalist economies could achieve full employment. They will not do so, though, because of the undue power that would then reside with workers, both individually and collectively. Belief in the capacity to create full employment is a consequence of the Keynesian theory that is at the heart of Kalecki's approach, in which effective demand determines the level of employment. This point will be taken up below. Second, once again, the approach does not really get to the heart of labour markets as distinct from other markets. Essentially, it is a general argument that supply cannot be allowed to become too powerful through guaranteed demand. But this applies to all markets, especially in the case of monopolies (addressed by competition policy) and those considered to be strategic for economic or other reasons, the military or the 'utilities'. As a result, such industries have been regulated or even nationalised.

What all of these explanations for the labour market have in common is a wish to explain its distinctiveness by reference to the nature of labour, its market or some combination of the two. This is why they are deficient since they range over specific instances of more general properties that do not sufficiently distinguish labour (markets) from others – it does not help much to ask whether labour is more or less fish-like or sold through more or less uncertain contracts. To be more direct, with an *ex post* exception that proves the rule suggested in the next but one sentence, what remains unexamined are the social relations that allow for the market or wage form of labour – as opposed to feudal or slave forms of labour for example. This is precluded by taking labour and/or markets themselves as the starting point. Underlying social relations can then only enter as an afterthought, as in the attempt of some political economy to suggest that labour markets are distinguished by their unique attachment to class, conflict and asymmetry in power.

FROM LABOUR MARKET TO VALUE PRODUCTION AND LABOUR MARKETS

Such factors should be taken as the starting point for understanding the labour market. And, for Marxist theory, doing so yields a remarkably simple yet rich answer as its implications are worked through to more complex and concrete outcomes. What distinguishes the labour market is that labour or, more exactly, labour power employed by capital is the source of value (and surplus value). From this, that the labour market is the form adopted by (capitalist) value relations, a whole series of propositions follow. First, and foremost, such relations are attached to particular class relations – those in which capital and labour confront one another. Because of the ownership of the means of production by capital, labourers can only gain access to work through the labour market. There are very few other realistic alternatives. Whilst there might appear to be a choice as market exchange is in some sense free by comparison with the coercion of feudalism or slavery, the possibilities are limited without denying the presence of, and possibilities created for, the self-employed, for example.

Second, with labour power employed by capital for the purpose of creating (surplus) value, the process of *production*, not exchange (or contracting), becomes paramount. The germane issue is how do capitalists appropriate surplus value. For Marx, the answer lies in extending the labour performed beyond that necessary to provide for the wage (see Chapters 1, 2 and 4). It is not a matter of stealing some of the worker's product, or part of the distribution of a net product after it has been produced. Marx examines exploitation, the production of surplus value, in terms of a fair exchange in the labour market. Both parties agree and no force is necessarily used, except when conflict between capital and labour becomes overt. But it is the goal of the capitalist to get sufficient work to leave surplus value over and beyond that required to match the value of the wage. This is done, according to Marx, either by persuading workers to labour longer or more intensively (what he terms absolute surplus value) or by increasing productivity of labour so that a given wage takes up less value and leaves more surplus value left even with a fixed working day and real wage (relative surplus value for Marx).

Third, both absolute and relative surplus value go far beyond simple logical categorisations of how capitalists appropriate more surplus value. Each has important implications for how the production process is controlled and evolves, and the conflicts it generates. And each is of greater or lesser significance in particular times and places. As capitalism becomes more advanced and the crudest forms of exploitation (low wages and long work) are eliminated, so relative surplus value comes to the fore, consolidated into a particular period or stage of capitalism, monopoly capitalism for Lenin. Here productivity increase is crucial and is underpinned by legislation to limit working hours, for example, or for social provision to allow for a healthy and skilled workforce.

Fourth, for Marx, the major systematic source of productivity increase is derived through increase in size of capital, through accumulation. In order to produce more commodities with a given quantity of labour, more raw materials are needed for processing, and sophisticated and large-scale fixed capital provides for these to be worked up into commodities.

Fifth, on this basis, Marx explains the presence of unemployment under capitalism. He perceives it to be both a condition for, and a consequence of, the accumulation of capital. As capitals grow and increase productivity, so they eliminate rivals whose workers are rendered unemployed. By the same token, those unemployed form what Marx calls a reserve army of labour on which expanding capitals can draw, as can those less advanced capitalists relying upon both low productivity and low wages as in sweatshops.

Sixth, it follows that employment and unemployment are heavily influenced by the scale and nature of capital accumulation, with workers gaining and losing work according to the accumulation and restructuring of capital. In this light, understanding the labour market in terms of equilibrium or natural rates of (un)employment becomes ludicrous – as if the nature of the ocean could be addressed in terms of the average sea level rather than its ebb and flow.

Seventh, Marx's theory of technical change is extremely sophisticated but is probably best known for the simple idea that capitalism divides production down into detailed tasks, displaces workers by machinery, and thereby deskills workers and reduces them to machine minders. But the increasingly complex nature of machinery, raw materials and production processes does itself also require that workers be reskilled. Consequently, the division of labour across skilled and unskilled (and what these are) depends upon how the accumulation of capital accommodates economies of scale and scope, quite apart from the scale and scope of capitalist production itself (for which products increasingly become subject to commercialisation as more and more activities are brought under the market umbrella).

Eighth, the restructuring both of capital and of labour markets is highly contingent on associated socio-economic processes, relations, and structures of broader scope. These concern urbanisation, indus-trialisation, demographic transition, social stratification and so on, with measures to promote the health, education and welfare of the working class in pursuit of productivity increase and to temper struggles for alternatives to the worst excesses of capitalism.

Ninth, taken together, these points lead to a simple conclusion.[4] There is no such thing as *the* labour market. For labour markets are highly differentiated from one another according to how the various factors involved interact with one another and are institu-tionalised within and across workplaces. Further, such differentiation is deepened by the possibilities for, and creation of, labour markets that are not necessarily or directly tied to capital and profitability – as in much of the state sector, some personal services and self-employment. In part, these promote the illusion that full employment is a possibility given a sufficiently strong pull on the effective demand lever (through state expenditure and/or employment). But the outcome cannot be freed from the rhythm and pace of capital accumulation. In short, on the basis of underlying value relations of production, outcomes for the labour markets are complex and need to be traced through a diverse set of interconnected factors embedded in the discussion of the preceding paragraphs.

FROM ECONOMIC TO SOCIAL REPRODUCTION

Not surprisingly, these factors are conditioned by, and based upon, a history of capitalism that is marked by a diverse range of struggles around the value relations that make up labour markets. Necessarily, at the forefront of such struggles are trade unions, the form of economic organisation of the working class in pursuit of its collective interests, however these might be defined and pursued. At its core, to the extent that the trade union movement aspires to a fair day's wage for a fair day's work (and for waged work itself or full employment), it cannot fundamentally challenge the wage relation and the class relations that underlie it.

The limitations of such a goal can be seen in a number of ways. First a fair day's wage has its counterpart in the sanctity of a fair profit for capital advanced, either as an ethical principle (reward for saving/investment/ownership) or as a pragmatic one (low profitability will squeeze out investment). Second, more generally, a fair day's wage leaves unattended those aspects of wage employment that were previously raised in terms of alienation – control of production in concept, organisation and execution and control of the products themselves. Third, fair wage gains are far from secure given the vagaries of capitalist competition and the cyclical crises of accumulation that bring depression and downward pressure on wages and employment. Fourth, gains for some workers in wages and employment can be at the expense of others, both individually and collectively, not least for example in competitive trade wars through protectionist measures.

Despite these reservations, wage struggle is not to be sniffed at and treated with suspicion for failing to challenge the system more fundamentally. First, trade union successes can result in real gains that can be generalised. Only in a narrow, static distributional understanding of wage struggles is conflict perceived to be zero-sum at best, with the ultimate sanction, resting with the capitalist, of job loss and closure should wages rise. Rather, wage gains are consistent with accumulation through more productive techniques and the elimination of sweatshops and extreme forms of exploitation – although restructuring of industry in this way is by no means guaranteed by wage rises.

Second, even in the economic arena, trade unions are not confined to wage struggles alone. They concern themselves to a greater or lesser extent with every condition of work, ranging across the spectrum covered by Marx's notion of alienated labour, as well as hours of work, its intensity, and breaks during work as well as for sickness and holidays, etc. Thus, workers struggle not only against capitalist control of the production process but also over the production conditions themselves, as with bargaining over new technology and the nature of workplace authority – what workers and managers may, or may not, do. Again there can be no guarantees over how far these struggles go and how successful and secure they are. But it is worth bearing in mind that they can even raise profitability. For workers have direct knowledge of the production process, and improvements in their conditions of work are not necessarily at the expense of productivity and profitability. Nonetheless, they can be resisted by capitalists as a matter of principle in view of their origins. For the principle of capitalist authority over production and property can take precedence over profitability, with Marx giving the example of employers resisting legislation to require guards to protect workers against injury from machinery. Once in place, such measures become, and appear to have been, uncontroversial.

Third, as is already apparent, in engaging in economic struggles, trade unions will often be drawn across the nebulous and shifting boundaries connecting economic with social reproduction. The wage, after all, is only the most immediate source of revenue for sustenance of the working-class family, whose capacity to provide able and skilled labour depends upon the range of 'services' that are now commonly thought of as constituting part and parcel of the welfare state, albeit unevenly by country and type of provision (housing, education, health, etc.). Initially, such non-wage struggles are conducted to limit the length of the working day and the conditions of employment of women and children.

Fourth, the rise of the welfare state does not derive exclusively from reform conceded for fear of working-class struggle and the drive for a more productive workforce. Nonetheless, trade unions have played a significant role in the formation and the continuing content of the welfare state, not least through trade unions within the state sectors themselves. Two different aspects are crucial. On the one hand, trade unions as economic organisations have the motive and the potential for participating in or creating political parties in pursuit of broader interests (although this has the dual effect of broadening the basis for struggle whilst displacing it from the point of production and from producers themselves). On the other hand, trade unions (and their political representatives) can successfully seek to 'decommodify' provision of welfare services, 'putting people before profit' in terms of how provision is made as well as ensuring that it is, in principle, free on demand. In these respects, embryonic socialist forms of production can already be seen to be evolving within capitalism, with the persistence of the wage form of labour but without the inevitable conditions attached to the capitalist production of surplus value.

FROM ECONOMISM TO SOCIALISM

No doubt the previous paragraphs present too rosy a picture of trade unions in practice in the absence of a downside that includes economism (the failure to move beyond immediate material issues, especially wages), the pursuit of sectional interests both nationally and sectorally, and a mixed response to new social movements concerned with sexism, racism and the environment. Defence workers, after all, depend upon the production of armaments and hence upon war, just as energy and car workers in a sense depend upon pollution. Yet the transition to socialism is to go through two phases according to Marx as he intriguingly and openly suggests in his *Critique of the Gotha Programme*. In the first phase, the bourgeois principle of reward according to work contributed continues to prevail:

> What we have to deal with here is a communist society, not as it has *developed* on its own foundations, but, on the contrary, as it *emerges* from capitalist society; which is thus in every respect, economically, morally and intellectually, still stamped with the birth-marks of the old society from whose womb it emerges. Accordingly the individual producer receives back from society – after the deductions have been made [for common fund for administration, health education and welfare] – exactly what he gives to it. (1875, p. 563)[5]

With some form of wage labour persisting in this first phase, it is essential that organised labour ensures that new, let alone old, forms of exploitation do not emerge, or re-emerge, this depending upon how production is controlled alongside the surplus labour itself. Only then, through a remarkable synthesis of critical commentary on capitalism and the prospects it promises, does Marx suggest:

> In a higher phase of communist society, after the enslaving subordination of individuals under division of labour, and therefore also the antithesis between mental and physical labour, has vanished, after labour has become not merely a means to live but has become itself the primary necessity of life, after the productive forces have also increased with the all-round development of the individual, and all the springs of co-operative wealth flow more abundantly – only then can the narrow horizon of bourgeois right be fully left behind and society inscribe on its banners: from each according to his ability, to each according to his needs. (p. 566)

REFERENCES AND FURTHER READING

Akerlof, G. and Yellen, J. (eds.) (1986) *Efficiency Wage Models of the Labour Market*. Cambridge: Cambridge University Press.

Bowles, S. and Gintis, H. (1990) 'Contested Exchange: New Microfoundations for the Political Economy of Capitalism', *Politics and Society* 18 (2), pp. 165–222.

Fine, B. (1985) 'On the Political Economy of Socialism: Theoretical Considerations with Reference to Non-European and European Experience', in D. Banerjee (ed.) *Marxian Theory and the Third World*. New Delhi/Beverly Hills/London: Sage Publications.

Fine, B. (1998) *Labour Market Theory: A Constructive Reassessment*. London: Routledge.

Kalecki, M. (1943) 'Political Aspects of Full Employment', *Political Quarterly* 14 (3), reproduced in Kalecki (1972).

Kalecki, M. (1972) *The Last Phase in the Transformation of Capitalism*. New York: Monthly Review Press.

Marx, K. (1875) *Critique of the Gotha Programme*, reproduced in *Selected Works in Two Volumes*. New York: International Publishers, 1933.

Solow, R. (1990) *The Labour Market as a Social Institution*. Oxford: Basil Blackwell.

Spencer, D. (2002) 'Shirking the Issue? Efficiency Wages, Work Discipline and Full Employment', *Review of Political Economy*, forthcoming.

NOTES

1. See Spencer (2002) for a critical account.
2. See Akerlof and Yellen (1986), for example.
3. See Bowles and Gintis (1990), for example.
4. For fuller discussion of what has preceded and this conclusion, see Fine (1998).
5. For extensive discussion, see Fine (1985).

6 Technological Change as Class Struggle

Les Levidow

What is technology? Apparently it works by helping people to solve problems. In our everyday lives, we commonly experience technology as hardware whose effects may be predictable or otherwise, beneficial or otherwise.

Yet political controversy has increasingly focused upon technology – e.g. automation, nuclear power, agrochemicals, biotechnology, electronic surveillance, etc. Protest has challenged not simply their undesirable side effects, but also their implicit purposes, i.e. how their designers define the problem to be solved. From where do these problem definitions come?

This chapter will make the following arguments:

- that technological change is inherently social, driven by class struggle, broadly understood;
- that prevalent technological designs embody strategies for exploiting labour, commoditising resources, and extending market relations;
- that these aims are both promoted and disguised by 'efficiency' rhetoric; and
- that alternative designs and social futures are possible, by defining differently the problem to be solved.

To develop those arguments, the chapter has the following structure: critical perspectives on technology; case studies of educational technology and high-tech seeds; and conclusions about technological change as class struggle. The analysis will draw on two key Marxist concepts – reification, whereby social relations take the form of relations between things; and fetishism, whereby a human quality takes the form of property of a thing.

TECHNOLOGY WORKS: EFFICIENCY FOR WHAT?

A common explanation for new technology is the need to increase efficiency (see Chapters 1 and 4). This begs fundamental questions:

efficiency for what purpose? under whose control? according to what account of progress? Let us survey some critical perspectives.

When Karl Marx analysed new technology in the Industrial Revolution, e.g. the steam engine and the self-acting mule, he argued, 'It would be possible to write a history of the inventions made since 1830 for the sole purpose of supplying capital with weapons against the revolt of the working class' (Marx 1976, p. 563). Such devices routinised, displaced and undermined craft labour, as a step towards extending the factory system. While wage labour already imposed a formal subordination of labour to capital, new technologies helped to achieve its real subordination, i.e. capitalist discipline over the content and pace of work, in order to maximise exploitation.

Since then, Marx's argument has been extended to another century-and-a-half of inventions. More recent technologies have been designed to reduce managerial dependence upon living labour – e.g. by replacing, managing or disciplining human labour. Their design codifies, embodies and appropriates specialist skills – not only those of shop-floor workers, but also those of professional staff and middle managers (e.g. Braverman 1976, Noble 1984, Robins and Webster 1985).

That class rationale can be illustrated by many examples from the 1980s onwards. The 'Information Technology Revolution' responded to the employers' problem that informal worker collectivities were exercising informal control over the labour process; they could keep down the work pace, could use their paid time for unofficial activities, and could even counterpose their own agendas. In the same period, the UK government expanded its nuclear power programme, while also automating the coal mines; such decisions responded to the problem that miners' revolts were catalysing class-wide solidarity. Such problem definitions influenced the R & D criteria for technological solutions. The state and private investors favoured new technological designs which served to fragment and discipline workers, while enhancing managerial authority over labour (see Chapter 17).

Given those aims, some critics have demanded to 'share the benefits', e.g. by increasing workers' wages or reducing their hours. Such proposals assume that the main benefits (indeed, purposes) are a quantifiable increase in material things which therefore could be allocated more fairly. Yet the benefits lie mainly in relationships of power.

Wage-labour discipline is a condition for its products to be fetishised as properties of things. As Karl Marx argued,

> the commodity reflects the social characteristics of men's own labour as objective characteristics of the products of labour themselves, as the socio-natural properties of these things ... I call this the fetishism which attaches itself to the products of labour as soon as they are produced as commodities ... To the producers, therefore, the social relations between their private labours *appear as what they are*, i.e. they do not appear as direct social relations between persons in their work, but rather as material relations between persons and social relations between things. (Marx 1976, pp. 164–6, emphasis in original)

Through commodity exchange, then, relationships between labourers appear as relations between things, i.e. between quantities of their labour. This appearance is no misperception or misrepresentation. Rather, it is the form of social relations which allows products to be exchanged as quantities of a homogenous quality – exchange value. Such relations '*appear as what they are*', in so far as human labour is subordinated to wage labour for commodity exchange.

This subordination can be enforced by technology, i.e. by dead labour. In capitalist machinery, 'the *social* characteristics of their labour come to confront the workers, so to speak, in a *capitalised* form; thus machinery is an instance of the way in which the visible products of labour take on the appearance of its masters'. Moreover, the forces of nature and science 'become separated from the skill and knowledge of the individual worker' (Marx 1976, p. 1055, emphasis in original).

Such perspectives were later elaborated, especially by the Frankfurt School of Critical Theory. According to Herbert Marcuse, modern-day 'efficiency' derives from the capitalist project of commoditisation. Heterogeneous qualities are homogenised into universally comparable ones, thus allowing a quantifiable output to be designed, measured, managed, increased, etc. Such homogenisation is promoted as a neutral technical tool, thus denying its own value-laden character. 'As universal functionalisation (which finds its economic expression in exchange value), it becomes the precondition of calculable *efficiency* – of universal efficiency' (Marcuse 1978, p. 205).

Such criteria are embedded in technology: 'Specific purposes and interests ... enter the very construction of the technical apparatus.'

Consequently, 'rational, "value-free" technology is the separation of man from the means of production and his subordination to technical efficiency and necessity – all this within the framework of private enterprise' (p. 222).

In more recent history, such strategies have been extended for producing novel commodities, for reproducing labour power amenable to capitalist work discipline, and for commoditising natural resources. Technological designs promote a commitment to ordering the world according to their own models of society and nature. A technology can 'work' only by creating the socio-natural conditions which its design takes for granted. At the same time, these inbuilt imperatives appear as properties of technology, rooted in its thing-like characteristics.

Those technological characteristics are not merely an appearance or rhetoric. Let us consider the social metaphors in 'smart bombs', 'intelligent machines', 'clean technology', 'gene banks', 'natural capital', etc. As particular human purposes are embedded into technology, those social qualities are fetishised as material or natural properties, while things acquire human-like qualities. These design choices confront us as a discovery about the nature of things. As quoted above, 'the visible products of labour take on the appearance of its masters' (Marx 1976, p. 1055).

In such ways, technological change both promotes and conceals class interests through future models of society. This role will be illustrated by two examples: Information and Communication Technologies (ICTs) in higher education; and genetically modified (GM) crops.

ICTs FOR CAPITALISING EDUCATION

Recent conflicts over educational values have intersected with designs for Information and Communication Technology (ICT). In the neoliberal project, marketisation imperatives are attributed to inherent qualities of ICT. According to the former Director-General of the WTO, Mike Moore, 'There are technical reasons for the acceleration of trade in services, especially in the area of information technology.' Through electronic transmission, local services have been 'transformed into internationally tradable products' such as education services, he argues.

Roles of ICTs

In such accounts, a political agenda is fetishised as an inherent property of technological progress. Electronic media generate an inevitable future, to which we must adapt through trade liberalisation. ICTs play several roles in this neoliberal agenda, especially for higher education.

First, according to the 'info-society' paradigm, the management, quality and speed of information become essential for economic competitiveness. ICT is dependent upon highly skilled labour; together they will be used in order to increase productivity and to provide new services, we are told. This imperative redefines the skills which higher education must provide.

Second, ICT facilitates the individualised and flexibilised learning which is required for modern workers. They must become individually responsible for managing their own human capital in the workplace. Through lifelong distance learning, for example, they will be able to recycle themselves at their own expense during their free time.

Third, teachers are identified as a problem. They have 'an insufficient understanding of the economic environment, business and the notion of profit', and their present role hinders 'internal searches for efficiency', according to industry lobbyists (e.g. ERT 1989). As a solution, ICT diminishes the role of teachers.

Fourth, universities are subjected to productivity criteria, as an imperative for their survival. They must package knowledge, deliver flexible education through ICT, provide adequate training for 'knowledge workers', and produce more of them at lower unit cost. While this scenario portrays universities as guiding social change, there is evidence of a reverse tendency: that they are becoming subordinate to corporate-style managerialism and income-maximisation. For neoliberal strategies, the real task is not to enhance skills but rather to control labour costs in the labour-intensive service sector, e.g. education (Garnham 2000).

Overall, neoliberal strategies for higher education have the following features. All constituencies are treated through business relationships. Educational efficiency, accountability and quality are redefined in accountancy terms; courses are recast as instructional commodities. Student–teacher relations are mediated by the consumption and production of things, e.g. software products, performance criteria, etc.

Prime agents of this agenda are the IMF and the World Bank, which elaborate the strategies of their paymasters in the dominant OECD countries. For several years the World Bank has been promoting a 'reform agenda' on higher education. Its key features are privatisation, deregulation and marketisation. A 1998 World Bank report identified the traditional university and its faculty members as inefficiencies to be rationalised. As the solution, 'Radical change, or restructuring, of an institution of higher education means either fewer and/or different faculty, professional staff, and support workers.'

Online solutions

In its future scenario, then, higher education would become less dependent upon teachers' skills. Students would become customers or clients. As the implicit aim, private investors would have greater opportunities to profit from state expenditure, while influencing the form and content of education. Business and university administrators would become the main partnership, redefining student–teacher relations. Although the World Bank agenda has little support among educators, some elements are already being implemented, e.g. in the guise of ICTs.

In North America many universities act not only as business partners, but also as businesses in themselves. They develop profit-making activities through university resources, casualised faculty and cheap student labour. In developing online educational technology, they aim to commodify and standardise education. Those aims have been resisted by students and teachers, e.g. by raising the slogan, 'the classroom versus the boardroom'.

In the name of increasing efficiency, North American universities have standardised course materials. Once lectures are submitted to administrators and posted on web pages, these materials can be merchandised to other universities. Better yet, the course writing can be outsourced on contract to non-university staff. By transferring control to administrators, the technology can be designed to discipline, deskill and displace teachers' labour. This approach changes the role of students, who become consumers of instructional commodities (Noble 1998).

The putative threat of market competition has been invoked as a serious threat to higher education in Europe. According to the national body of university executives, Universities UK, the solution is to abolish borders between the university and business, as well as those between domestic and international 'markets' for educational goods. They describe the university as already a business, albeit a deficient one which must be corrected according to corporate principles. Their chief celebrates changes in undergraduate delivery:

from a 'just-in-case' general intellectual training, to a more flexible 'just-in-time' ethos, and then to 'just-for-you' forms of learning.

In particular, they promote internet-based delivery as a key means of becoming a 'borderless business'. According to sponsors of the electronic university, 'The project is designed to give UK higher education the capacity to compete globally with the major virtual and corporate universities being developed in the United States and elsewhere.' The preliminary business model 'recommended that pedagogic support should be embedded within learning materials, and that supplementary online support might be negotiated for individual students at a price'. This proposal generated internal debate about what types of social interaction must be designed into the product in order to find customers. By standardising courses and reducing dependence on professional skills, the electronic university seeks ways to commodify student–teacher relations.

In sum, the neoliberal project cites putative imperatives of technological forces, whose design in turn reshapes higher education in the image of a marketplace. Through ICT, marketisation can take the apparently neutral form of greater student access, flexible delivery, efficiency, etc. As education is capitalised, student–teacher relations are reified as relations between things, e.g. between consumers and providers of instructional software (Levidow 2001).

Rather than 'inefficiency', we can define marketisation as the problem. It threatens academic freedom, internal democracy and the university's scope for critical analysis. ICT could be used instead to promote critical citizenship, to link student networks, to create 'virtual communities' of interest in controversial issues, and to circulate debate on critical perspectives. Technology could be designed differently to facilitate alternative futures.

SEEDS FOR CAPITALISING NATURE

Seeds have been a site of struggle over the entire agro-food chain. Given the inherent reproducibility and variability of seeds, their natural characteristics have provided an opportunity for farmers to improve varieties through selective breeding, while developing their skills in cultivation methods. Capitalist strategies have attacked that independence by uncoupling seeds from farmers' control, thus capitalising natural resources into commodities. Even by the mid-nineteenth century, noted Marx (1973, p. 527), 'Agriculture no longer finds the natural conditions of its own production within itself, naturally arisen, spontaneous, and ready to hand, but [rather] these exist as an independent industry separate from it.'

A major step in capitalising nature was hybrid seeds, which could not breed true, so that farmers had to buy them anew each season. Citing the supposed benefits of 'hybrid vigour', agricultural research prioritised such varieties, rather than improve open-pollinated varieties. For seeds which could still be bred under farmers' control, the proprietary gap was filled by a series of laws restricting farmers' rights to sell or even reproduce their own grain for seed. They have also been locked into dependency through other purchased inputs, grain contracts, debt, etc., such that the farmer has become little more than a propertied labourer, despite owning the land (Kloppenburg 1988).

High-response varieties

Such dependency was extended to the South through the Green Revolution. So-called 'high-yielding varieties' (HYVs) were really 'high-response varieties'. Designed for more intensive cultivation methods, their higher yields depended upon agrochemicals, irrigation and other purchased inputs. That entire system was fetishised as an inherent property of a discovery, the HYV.

The use of HYVs substantially increased grain yields of wheat and rice in India, yet this increase counted as greater efficiency only by measuring a single commodity, while ignoring previous beneficial practices. Higher grain yield meant less straw, used locally as animal feed. Previously many farmers had done intercropping – e.g. sorghum and wheat with pulses – which helped to renew soil fertility, while providing other nutrients. Those benefits were lost in the switch to HYVs. More generally, land use shifted away from cultivating oilseeds and pulses, which had been a cheap protein source – 'the poor person's meat'. Eventually India had a shortage of oilseeds and pulses, which had to be obtained through imports.

Moreover, HYVs favoured those farmers who could obtain loans for the purchased inputs. Financial dependency and market competition drove many into debt, even out of business, leading some to commit suicide. Landless peasants became wage labourers for the successful farmers or migrated to cities. Some moved to live or work near the Union Carbide plant at Bhopal, which supplied agrochemicals for the HYVs and was the site of the poison gas disaster (Shiva 1991).

Those outcomes logically followed from the agribusiness agenda which guided the HYV research, largely funded by the Rockefeller Institute. According to its chief, 'agriculture is a business and, to be successful, it must be managed in a business-like fashion'. As such language indicates, the Green Revolution redefined agricultural efficiency in terms of calculable commodities, while devaluing any resources or benefits which did not fit such a model.

Agbiotech

That commoditisation agenda has been extended by agricultural biotechnology, which originated from the 1930s science of molecular biology. This science reconceptualised 'life' in physico-chemical terms: through a computer metaphor, DNA became coded 'information' which could be freely transferred across the species barrier. 'As technology controlled by capital, it is a specific mode of the appropriation of living nature – literally capitalising life' (Yoxen 1981).

This technological trajectory has intersected with a wider debate over sustainable agriculture. At issue is how to diagnose and remedy the systemic hazards of intensive monoculture, e.g. agrochemical pollution, pest epidemics, pest resistance, etc. Agbiotech attributes these problems to deficient seeds, which must be corrected by editing their genetic information, thus making agriculture more efficient and clean.

Through social metaphors, nature is recast in the image of biotechnology, while human qualities are fetishised as properties of 'smart seeds', i.e. genetically modified (GM) crops. In addition to the computer-code metaphor, biotechnologists speak of a clean surgical precision, e.g. redesigning seeds to attack pests or to withstand herbicides, which are often sold by the same company. Some also speak of 'value-added genetics', i.e. searching for genetic changes which can enhance the market value of agricultural products. By projecting capitalist criteria onto nature, design choices are fetishised as the discovery of natural properties (Levidow 1996).

Moreover, farmers' socio-economic dependency on purchased inputs is reified as a relation between things – e.g. as a relation between crops and external threats. Ironically, such dependency is portrayed as liberation from natural threats to crops. Although farmers may still have the free choice to buy non-GM seeds, the GM option can become an imperative, given the promise of greater efficiency and the consequent threat of market competition. For example, some crops are redesigned as interchangeable, flexibly sourced raw materials. Others are designed to favour large planta-tions over small-scale farmers. In various ways, the consequent pressures benefit mainly the agro-food industry (Hobbelink 1991).

Biopiracy

GM techniques have also been used as a symbolic instrument to privatise seeds. According to advocates of greater patent rights, GM crops are inventions, involving a significant contribution by scientists. According to opponents, however, such products are dis-coveries (or simulations) of common resources which have been already selected and cultivated by farmers over many generations.

To qualify for patent rights, a product or process must have con-tributed an 'inventive step'. The USA and the European Union have interpreted that criterion so as to accept broad patent claims on GM crops. Patents have encompassed substances derived from plants tra-ditionally cultivated in developing countries, e.g. pesticidal agents from the neem tree. Even some non-GM seeds have been subjected to royalty payments by farmers in Southern countries, thus threat-ening their livelihoods. The US government has sought to extend its patent criteria to other countries, by using the TRIPS (Trade-Related Aspects of Intellectual Property Rights) rules under the WTO.

Amid that conflict, the term 'biopiracy' has acquired two opposite meanings. For advocates of greater patent rights, 'biopiracy' means violating those rights, e.g. by using patented materials without a licence agreement or without paying royalties. For opponents of such rights, 'biopiracy' means the patents themselves – e.g. on the grounds that plant material should remain freely reproducible as a common resource, or that Southern farmers should be reimbursed for their plant-breeding skills.

Those stakes have generated fierce protests against GM crops, as well as a search for alternatives. In 1999 Indian farmers 'cremated' field trials of Monsanto's GM insecticidal cotton. In previous years, many of their fellow farmers had been abandoning mixed farming systems in favour of cotton monoculture based on hybrid seeds, thus intensifying their dependence upon purchased seeds, which sometimes led to crop failures. Towards an alternative future, their organisations encouraged farmers to resow their own seeds, to preserve diverse cultivars in the agricultural field, and to develop independent methods of crop protection. Likewise, the Sem Terra (landless) movement in Brazil has been developing organic cultivation methods, as an alternative both to chemical-intensive ones and to GM crops (Branford and Rocha 2002, 2003).

CONCLUSION: REDESIGNING TECHNOLOGY

As this chapter has shown, technological design is inherently social. Like history in general, technological change is driven by class struggle. Often its design responds to the employers' problem that workers resist subordination to capitalist labour discipline, e.g. through their informal collectivities and skills. The prevalent designs embody strategies for exploiting labour, commoditising resources, and extending market relations to more social activities.

These tendencies are both promoted and disguised by 'efficiency' rhetoric, which values human activities only as quantifiable things. As such aims guide R & D, the resultant technology tends to acquire human qualities, which become fetishised as properties of things. Technological imperatives and solutions appear to arise from the natural order. Social relations become reified as relations between things, e.g. between smart seeds and external natural threats, or between consumers and providers of instructional software.

Through various forms of resistance, however, people attempt to dereify those relations: they seek to reappropriate their collective power and skills, while expressing needs beyond market relations. By defining differently the problem to be solved, alternative designs become possible (see, for example, Sclove 1995). Rather than a discovery of natural properties, technological change can be recast as diverse human choices for the future of society.

REFERENCES AND FURTHER READING

Branford, S. and Rocha, J. (2002) *Cutting the Wire: The Story of the Landless Movement in Brazil*. London: Latin American Bureau.
Branford, S. and Rocha, J. (2003) 'Another Modernisation is Possible: From Land Conquest to Agro-ecological Experiment', *Science as Culture* 12(1).
Braverman, H. (1976) *Labour and Monopoly Capital*. New York: Monthly Review Press.
ERT (1989) *Education and Competence in Europe*. Brussels: European Round Table of Industrialists (www.ert.be).
Garnham, N. (2000) '"Information Society" as Theory or Ideology', *Information, Communication and Society* 3(2), pp. 139–52.
Hobbelink, H. (1991) *Biotechnology and the Future of World Agriculture*. London: Zed Books.
Kloppenburg, J. (1988) *First the Seed: The Political Economy of Plant Biotechnology, 1492–2000*. Cambridge: Cambridge University Press.

Levidow, L. (1996) 'Simulating Mother Nature, Industrializing Agriculture', in G. Robertson et al. (eds.) *FutureNatural: Nature, Science, Culture*. London: Routledge.

Levidow, L. (2001) 'Marketizing Higher Education: Neoliberal Strategies and Counter-Strategies', *Education and Social Justice* 3(2), pp. 12–23, available from www.trentham-books.co.uk, www.commoner.org.uk/03levidow.pdf, and from K. Robins and F. Webster (eds.) *The Virtual University? Information, Markets and Managements*. Oxford: Oxford University Press, 2002.

Marcuse, H. (1978) 'Industrialization and Capitalism in the Work of Max Weber', in *Negations: Essays in Critical Theory*. Boston: Beacon Press and London: Free Association Books 1988.

Marx, K. (1973) *Grundrisse*. London: Penguin.

Marx, K. (1976) *Capital*, vol. 1, especially 'The Fetishism of the Commodity and Its Secret', pp. 163–77. London: Penguin.

Noble, D. (1984) *Forces of Production: A Social History of Industrial Automation*. New York: Alfred Knopf.

Noble, D. (1998) 'Digital Diploma Mills: the Automation of Higher Education', *Monthly Review* 49(9), pp. 38–52; also in *Science as Culture* 7(3), pp. 355–68; other material available at thecity.sfsu.edu/~eisman/digital.diplomas.html and www.communication.ucsd.edu/dl/ddm2.html.

Robins, K. and Webster, F. (1985) 'New Technology and the Critique of Political Economy', in L. Levidow and R.M. Young (eds.) *Science, Technology and the Labour Process: Marxist Studies*, vol. 2. London: Free Association Books.

Sclove, R. (1995) *Democracy and Technology*. New York: Guilford Publications.

Shiva, V. (1991) *The Violence of the Green Revolution*. London: Zed Books/Third World Network.

Yoxen, E. (1981) 'Life as a Productive Force: Capitalizing the Science and Technology of Molecular Biology', in L. Levidow and R.M. Young (eds.) *Science, Technology and the Labour Process: Marxist Studies*, vol. 1. London: CSE Books/Atlantic Highlands: Humanities Press; reissued by Free Association Books, 1983.

7 Capitalism, Nature and the Class Struggle[1]

Paul Burkett

There was a time when those voicing concerns about environmental crises and the quality of our relationship with nature were quickly dismissed as alarmists, doomsayers, and starry-eyed romantics standing in the way of human progress. No more. Now the media are filled with reports on global warming, the breakdown of the ozone layer, destruction of rainforests, reduced species diversity, depletion of non-renewable energy sources, and the build-up of carcinogens and other toxins in our air, water and food. Even the word 'development' is as likely to carry images of urban sprawl gouging, poisoning, and eroding the land as of a comfortable and secure human existence.

The demand for a sustainable and healthy relationship with nature is central to the growing worldwide rebellion against the current system of development. Yet, this challenge has been weakened by an inability to uncover and describe clearly the systemic roots of environmental crises. The popular movements are increasingly led by those who blame ecological problems (and much else besides) on particular institutions in our economic system, especially transnational corporations and banks along with the IMF, World Bank and WTO. But attempts to analyse these institutions and their activities using the category of 'globalisation' seem to beg the question as to what social relations are being globalised. What kind of socio-economic system creates such powers as these global institutions wield?

Similarly, ecological economists have pointed out the failure of mainstream economics (especially neoclassical growth theory) to recognise the irreducible role of limited natural resources in human production. But the social relations that cause the current system to undervalue these resources are still not critically addressed. In this regard, production remains just as much a 'black box' for ecological economists as for mainstream economics.[2] Interestingly, the technical perspectives of economists complement the cultural

106

views of 'deep ecologists' who blame environmental crises on human-centred thinking and industrialisation as such – not the specific human–natural relations produced by specific socio-economic systems.

This failure to specify the social roots of environmental crisis can only hamper our political and intellectual challenges to the current form of development. To be successful, we need an accurate conception of the system we are fighting against – and that system is *capitalism*. Previous economic systems had their own environmental problems, but it is capitalism that is at the root of contemporary environmental crises and which has brought the human race to the brink of a planetary catastrophe. What are the fundamental relations that distinguish capitalism from other economic systems?

WEALTH, CAPITAL AND CAPITALISM

Any conception of capitalism hinges on prior conceptions of wealth and capital. Following Marx, I define wealth (use value) as anything that contributes to human life. Both nature and human labour are necessary sources of wealth, but the ways in which nature and labour are combined depend on the social relations in and through which human beings appropriate nature and transform it into humanly useful forms. In this sense, wealth is simply human development itself, a process that always involves a socially structured material 'metabolism' between people and nature. While recognising nature's contribution to human production, this conception of wealth also encompasses nature as an eternal condition of human life in all its material, intellectual, aesthetic, and even spiritual dimensions. Nowadays, living in a system that identifies 'wealth' with monetary and financial assets, it is easy to lose sight of this original, *human–natural* meaning of wealth.

Capital is the advancement of money to obtain more money, or M ... M'. *Capital accumulation* refers to the reinvestment of all or part of the monetary surplus so obtained (M'–M) to obtain yet more money (see Chapter 1). Such monetary accumulation has, of course, been taking place for quite a long time. From the bible we learn that merchants accumulated money through trade in goods (including slaves) and money-lenders became rich long before the birth of Christ. Commodity trade and money-lending can occur on the basis of a wide variety of social relations of production. All that is required is that different production units (households, villages, slave plantations, or feudal estates) desire to sell at least some surplus products (even a small amount above those produced for own use) to obtain other goods and services from other production units.

What distinguishes *capitalism*, as an economic system in its own right, is that money-making becomes the driving, overriding goal of production itself – a goal that competition inexorably presses upon the human producers. For this to happen, the producers must be socially separated from necessary conditions of production, above all the land but also the tools and machines used in production. Once this separation occurs, the only way that the economically disenfranchised producers can obtain needed means of consumption is by purchasing them as commodities using the wages obtained by working for the capitalist who owns the necessary conditions of production.[3] Under capitalism, in other words, labour and nature – the two basic sources of wealth – are first socially separated and then united only in production managed in line with the goal of money-making.

From this perspective, the phenomena commonly grouped under the category of economic globalisation comprise merely the latest phase of capitalism and capital accumulation on a global scale. Historically, the development of capitalism in Western Europe and North America was itself contingent on the establishment and growth of world markets in goods, money, and labour power; the plunder of natural and human wealth from colonised areas; and the establishment of an unequal division of labour between the developed capitalist centres and the colonies.[4] The agencies normally associated with contemporary globalisation are either institutional vehicles of capital (transnational corporations and banks) or fulfil necessary legal and financial functions for globalised capital accumulation (World Bank, IMF, and WTO). To focus our critical attention on these agencies rather than the social relations they embody is to evade a crucial question: How do capitalist relations influence people–nature relations?

CAPITALISM AND NATURE

In its own way, capitalism has been enormously productive of wealth compared to previous economic systems. By socially separating labour from natural conditions and developing both under the spur of competitive money-making, capitalism makes the development of production much more dynamic and transformative than, say, feudalism. In the latter system, the landlords basically viewed the land and everything on it (including the peasants) as given conditions of production. By contrast, capital treats labour and its natural conditions as mere vehicles of monetary accumulation, to be appropriated, divided, recombined, reshaped, and discarded as needs be in the never-ending pursuit of increased capital values. This difference accounts for the rapid development of the productive powers of labour and nature under capitalism compared to feudalism.

Unfortunately, the same factors that make capitalism a dynamic system of wealth creation render its production profoundly exploitative of and damaging to both labour and nature. The alienation of labour from natural conditions is manifested, for example, in capitalism's reduction of 'value' to the amount of social labour time directly objectified in commodities. For capitalism as a total system, the only source of *surplus* value or profit is the *surplus labour time* of the direct producers. Capital accumulation is thus much more directly dependent on the exploitation of *workers' time* than was the case with previous systems in which labour was not as fully separated from the conditions of production. Along with its adverse effects on workers, this specifically capitalist reduction of 'value' or 'wealth' to commodified labour time abstracts from nature's eternal contribution to real wealth – as if people can be separated from nature.

Naturally, capitalists do not directly observe value as commodified labour time in the abstract. For them, the pursuit of 'value' appears as the competitive process of money-making (M ... M') by any and all available means.[5] Considered formally, this process has several crucial characteristics. First, the goal of capital accumulation, namely, money, is a qualitatively homogenous entity. Its main quality *is* that of being pure quantity. Second, the basic monetary units of capital accumulation are, for all intents and purposes, perfectly divisible and mobile. Third, there is no apparent quantitative limit to the capital accumulation process. No matter how much money one has, there is always a larger amount that one can strive for. Importantly, none of these three features is shared by human–natural wealth. Considered as ecological systems, human communities and their natural environments are qualitatively variegated, interconnected (that is, imperfectly divisible), locationally specific, and quantitatively limited. These characteristics place definite limits on the appropriation of human–natural wealth – limits that, if exceeded, result in a qualitative deterioration of this wealth. Monetary accumulation, by not recognising the basic characteristics of real wealth, in effect denies all human–natural limits to wealth production. By placing monetary accumulation in command of production, capitalism guarantees that these limits will be over-stretched unless and insofar as society imposes external restraints on the capitalistic appropriation of labour and nature. This is as true for the limits to exploitation of nature as it is for the limits to the length and intensity of wage-labour time that human beings can endure.

CAPITALISM AND ENVIRONMENTAL CRISES

The contradiction between capital accumulation and natural wealth is manifested in two distinct kinds of environmental crisis: crises of capital accumulation and crises of human development. In the first type of crisis, the profitability of production is disrupted by the failure to respect the natural conditions of wealth production. One form this takes is shortages of the materials and non-renewable energy sources used to produce commodities. Agricultural production of raw materials, for example, is bound up with natural conditions which place certain limits on the amount that can be produced within any given time period – even with the application of human labour and technology. The same goes for forest products and other animate and plant materials gathered from nature. Some raw materials (minerals, metals, oil) as well as energy sources (coal, and again oil) are, in practical terms, completely non-renewable, that is, non-reproducible. When production is driven by competitive monetary accumulation, with its non-recognition of all quantitative limits, it is inevitable that the demand for materials and energy will periodically outstrip supplies. The resulting rise in material and energy prices, and absolute shortages of these crucial inputs, may produce disturbances and breakdowns in the profitable production and sale of commodities. This type of crisis is familiar from recent recessions in the advanced capitalist countries triggered by increases in the price of oil. In the mid-nineteenth century, periodic shortages of cotton had a similarly disruptive effect on capital accumulation.[6]

Capital accumulation may also be disrupted by the environmental disasters that it periodically produces. Capital's reduction of human–natural wealth to a vehicle of money-making, and its wanton disregard for the ecological qualities of this wealth, necessarily result in episodes of mass destruction of human beings and their natural conditions due to 'accidental' explosions, fires, nuclear meltdowns, chemical releases, etc. Such crises may themselves produce temporary shortages of exploitable labour power and materials, and in extreme cases may even put many production facilities out of commission.

Depending on the outcomes of political conflicts (e.g. class action lawsuits), these disasters can also impose 'clean-up costs' on capitalists – further disturbing profitable accumulation.

Nonetheless, capital accumulation can and will continue as long as capital can access a supply of wage labourers, and as long as these wage labourers can produce vendible commodities. Apart from these minimal requirements, capitalism is in no way dependent on particular natural conditions. It is an extreme form of what Gary Snyder has termed 'biosphere culture', that is, a society that 'spreads its economic support system out far enough that it can afford to wreck one ecosystem, and keep moving on'.[7] For capitalism, in short, there are no permanent environmental crises short of the total extinction of human life. Indeed, capital often profits from its own damaging effects on natural wealth – as when it produces and sells air conditioners and oxygen masks so people can 'live with' urban air pollution, and more generally when it enters the highly profitable and ever-expanding pollution control and waste management industries.

Capitalism's ability to survive and even prosper (in its own money-making terms) despite its plunder and vitiation of natural wealth makes it all the more crucial that we avoid identifying 'environmental crises' with crises *for capital accumulation*. We must recognise a second type of environmental crisis produced by capitalism: a crisis in the *quality* of human–natural wealth, i.e. in the *conditions of human development*. This second kind of crisis itself involves two historical dynamics operating both within countries and on a global scale. The first is capitalism's extreme division between city and country. Capitalist development concentrates both population and non-agricultural production in urban areas, whose material effluvia (human bodily waste and other forms of refuse), rather than being reproductively returned to the soil, instead pollute the cities and their hinterlands. Meanwhile, agriculture, when likewise harnessed to the quantitatively unlimited goal of monetary accumulation, becomes a 'factory farm' system reliant on ever growing 'fixes' of the fossil-fuel-intensive machinery, fertilisers and pesticides provided by urban factories. The resulting destruction of natural soil fertility and rural ecosystems is accentuated by the failure to recycle the animal waste produced by concentrated 'feedlot' facilities for raising livestock. In this way, capitalism's division of city and country creates an unhealthy and unsustainable circulation of matter between human civilisation and its natural environment.[8]

The second dynamic is a product of capitalism's latest phase of 'globalisation', i.e. of the further development of capitalist industry and the city/country division on a global scale. Since the Second World War, the scientific utilisation of wealth in the service of profit has created a new, more radical break between production and the natural conditions of human life. The combination of absolute growth of production and material/energy throughput on the one hand, and the development of non-biodegradable and downright poisonous forms of production on the other, has caused global capitalism to enter what John Bellamy Foster terms 'a new stage of planetary crisis in which economic activities begin to affect in entirely new ways the basic conditions of life on earth'.[9] Among the symptoms of this planetary crisis dynamic are global warming, ozone depletion, accelerated extinction of plant and animate species, and the proliferation of cancerous disease.

These dynamics clearly manifest capital's lack of concern for the qualitative variety, interconnection, locational uniqueness, and quantitative limits of natural wealth. That the planetary crisis stage corresponds to huge increases in monetary 'wealth' as measured by GDP, profits and financial asset values strongly suggests that capitalism has reached its historical limits as a system of production and human development.

MARKET-BASED ENVIRONMENTAL POLICIES

Capitalism's supporters deny the need for a basic change in economic relations to deal with environmental problems. They assert that the market system, by pricing natural resources, can limit the use and destruction of these resources – as long as clear private property rights are assigned to them. They also argue that, for those environmental problems involving privately unpriced 'external effects', or 'common pool resources' that are by their nature collective rather than private, the government can set up a system of artificial prices or rents using taxes, subsidies and tradable pollution permits.[10] But this argument assumes what needs to be shown, namely, that human–natural wealth can be adequately registered in profit-seeking market transactions. Market-based environmental policies do nothing to counteract the anti-ecological characteristics of money and capital. In fact, by assigning prices to natural resources (including 'ecosystem services') such policies legitimise the reduction of human–natural wealth to a means of money-making.

Whether set privately or publicly, market prices or rents do not effectively limit the depletion and destruction of natural resources. As we know from the fishing and oil industries, a higher price for a scarce resource is as likely to increase as to reduce its depletion by profit-seeking enterprises. Taxing the use of a resource will, like raising its price, tend to accelerate the depletion of substitute resources. If the tax is not globally enforced, it may accelerate depletion of the target resource in non-taxed countries and regions. Besides, taxes offer, at the very best, a highly imperfect *micro* instrument for achieving *macro* goals arrived at *prior to* market processes. This has become quite evident in recent negotiations surrounding biospheric problems like global warming and ozone depletion. Market prices and taxes also reinforce the arbitrary division of human–natural systems implicit in the money form of value, and this makes them a limp instrument for alleviating the ecological impacts of an increasingly intensive and globalised production system driven by the boundless goal of monetary accumulation. Given that the competitive search for resource rents is an important contributor to the planetary crisis produced by capitalism, it is difficult to see market prices and taxes as part of a viable solution unless we define sustainability in terms of continued growth of GDP.

THE CLASS STRUGGLE OVER NATURE

Capitalism's profit-driven development of production has created a global system of intensive human appropriation and processing of nature. This system calls out for a collective and democratic form of regulation, not only to protect the natural wealth that remains but also to restructure production in more healthy and sustainable directions. But the basic social relations of capitalism, specifically the separation of workers from conditions of production and the management of production by anarchically competing capitalists, stand in the way of such ecological regulation.

Nonetheless, we should not totally reject all legacies of capitalism in a vain search for a 'more natural' or 'less industrial' way of life. An important by-product of capitalist development has been a tremendous increase in human knowledge about natural wealth. But this knowledge has mainly been put to use on the micro level by competing enterprises producing for a profit, the result being growing environmental problems on the macro level. This paradox of macro irrationality alongside ever greater rationality in the micro-engineering of production accurately reflects the alienation of working people from the natural and social conditions of production. If this social separation can be overcome, the scientific knowledge developed under capitalism can be augmented and put to better use on macro *and* micro levels.[11]

A related paradox is that, by socially separating workers from natural conditions of production, capitalism makes 'the environment' a legitimate issue in its own right to a much greater degree than under previous systems like feudalism, in which the workers were not formally separated from the land. True, capitalism's 'separation' of environmental issues from workplace issues is itself a form of alienation, as is shown by the 'jobs versus environment trade-off'. But, from a class perspective, there is still a positive potential here: as capital tries to reduce labour and nature to means of money-making, it is in constant tension with the struggles of workers to improve their material and social conditions. In these struggles, workers and their communities are forced to confront and overcome capitalism's separation of workplace and environmental concerns.[12]

These struggles are of two broad types. First, workers may look beyond the sphere of capitalist production for less money-driven forms of human existence and fulfilment, in the process creating new cultural, political, and economic relations that challenge capital's dominance over nature *and* society. Second, they may resist capital's dominance in the sphere of wage labour itself, not only by demanding wage increases, safer and less burdensome work procedures, and reduced work times, but also by struggling for more co-operative and democratic forms of ownership and management of industry. Although they overlap, combine, and even clash in complex ways, the ultimate success of either mode of struggle in displacing the power of capital arguably depends on the other. And both modes of struggle require worker co-operation, democracy and solidarity as opposed to capital's reliance on exploitation, hierarchy, and atomistic competition. The former set of values is much more likely to produce sustainable and humanly sensible environmental policies. In short, workers' struggles (inside and outside the workplace) contain a powerful pro-ecological potential insofar as they contest all forms of money-driven exploitation of labour and nature.

To engage in the class struggle on the side of working people is to reject the search for technical and market-based fixes for environmental problems. The kinds of changes needed are those that help clean up the environment and move production in ecological directions *through increased worker and community control over production*. The power of money, capital and competition must yield to a conscious and collective socialisation of labour and nature. Subsidies for solar power to replace non-renewable energy sources? Definitely, but in ways that enhance worker-community control and reduce the role of profit-driven decisions and unplanned regulation by market forces. Reductions in work time? Yes, but as a means to increase the quantity and quality of citizen participation in economic and political decision-making, including the management of the community's natural and social conditions. Monetary accounting should be replaced by ecological accounting, i.e. by direct measures of the material and energy throughput from production and the ecological impacts thereof – with due regard for the inherent uncertainties involved. But, insofar as it guides production decisions, this accounting must adequately register the multiple dimensions of human–natural wealth, including concerns for other species. The only way this will happen is if workers and communities have a clear grasp of, and real participation in, the accounting process – so that conflicts can be worked out in the open, not smothered under the false balance sheets of monetary 'wealth'. Given the current dominance of money in capitalist society, getting ourselves to think in terms of such a collective non-monetary accounting promises to be the most difficult struggle of all.

REFERENCES AND FURTHER READING

Altvater, A. (1993) *The Future of the Market*. London: Verso.
Brewer, A. (1990) *Marxist Theories of Imperialism* (2nd edn.) London: Routledge.
Burkett, P. (1999) *Marx and Nature*. New York: St. Martin's Press.
Capital and Class 72, Autumn 2000 (special issue on environmental policy).
Capitalism, Nature, Socialism, any issue.
Chattopadhyay, P. (1994) *The Marxian Concept of Capital and the Soviet Experience*. Westport, Conn.: Praeger.
Daly, H.E. (1991) *Steady-State Economics*. Washington, D.C.: Island Press.
England, R. (2000) 'Natural Capital and the Theory of Economic Growth', *Ecological Economics* 34(3), September, pp. 425–31.
Foster, J.B. (1999) *The Vulnerable Planet*. New York: Monthly Review Press.
Foster, J.B. (2000) *Marx's Ecology*. New York: Monthly Review Press.
Heal, G. (2000) *Nature and the Marketplace*. Washington, D.C.: Island Press.
Hughes, J. (2000) *Ecology and Historical Materialism*. Cambridge: Cambridge University Press.
Jackson, W. (1994) *Becoming Native to This Place*. Lexington: University Press of Kentucky.
Lebowitz, M. (1992) *Beyond Capital*. New York: St. Martin's Press.
Marx, K. (1967) *Capital*, vol. 1. New York: International Publishers.

Perelman, M. (1987) *Marx's Crises Theory*. New York: Praeger.
Rosdolsky, R. (1977) *The Making of Marx's 'Capital'*. London: Pluto Press.
Snyder, G. (1977) *The Old Ways*. San Francisco: City Lights Books.

NOTES

1. The author thanks Don Richards and Alfredo Saad-Filho for useful comments on a prior draft.
2. See Daly (1991) and England (2000).
3. There is no presumption here that 'the capitalist' must be a private individual or corporation. It could be the state that owns and controls necessary conditions of production, which are then distributed among competing state enterprises. See Chattopadhyay (1994).
4. Brewer (1990), and references therein.
5. Marx (1967, part 1) showed how value in the sense of commodified labour time necessarily takes on the form of money; see also Rosdolsky (1977, ch. 5).
6. Such disruptions may involve increases in the *value* of materials (as scarcity exhibits itself in an increased labour time required to produce materials of given quality), rising *rents* to landowners, or both. See Altvater (1993, ch. 5), Burkett (1999, pp. 108–9) and Perelman (1987).
7. Snyder (1977, p. 21).
8. Foster (2000, ch. 5), Burkett (1999, pp. 119–28) and Jackson (1994).
9. Foster (1999, p. 108).
10. Heal (2000).
11. Under capitalism's money-driven production, certain kinds of knowledge about nature's ways, and about the requirements of ecological balance in human production, have been socially marginalised and in some cases forgotten despite the general growth of natural scientific knowledge (Jackson 1994, Snyder 1977). An important task of any anti-capitalist ecological movement, and of post-capitalist society, is to recover and enhance these ecological insights.
12. Lebowitz (1992).

Part II
Global Capitalism

8 The History of Capitalism

Michael Perelman

BACKGROUND

The history of capitalism is a contentious matter. Just as nobody has been able to pinpoint satisfactorily the moment at which life begins, some people are even able to find what they consider to be elements of capitalism in ancient society. For example, archaeologists have found records of business transactions in ancient Mesopotamian society.

In fact, some economists go so far as to report evidence of the existence of capitalism in the animal kingdom. Because they consider capitalism to be an innate extension of human nature, these economists have reported aspects of this same 'human' nature in rats (Battalio, Kagel and McDonald 1985). Adam Smith took a different position, suggesting that this 'instinct' for exchange was a defining quality of *Homo sapiens*. He asserted, 'Nobody ever saw a dog make a fair and deliberate exchange of one bone for another with another dog' (Smith 1776, I.ii.2, p. 26).

In reality, the defining characteristic of a capitalist society is the dominant form of social relations. The fundamental capitalist social relation is the relation between wage labour and capital. Within this arrangement, capital, represented by an individual employer or a firm, hires people with the intention of profiting from their work (see Chapter 1).

Notice the reference to the dominant form of social relations. The existence of an instance of a particular type of behaviour, an institution, or even some capitalist-like transactions, does not constitute evidence of a capitalist society. I can personally attest to this statement because I work in an institution fully steeped in feudal traditions; namely, a university, complete with an archaic administrative structure and apprenticeship rituals. On graduation day, the costumes that the faculty wear provide striking evidence of the feudal nature of this institution.

Nonetheless, nobody would think to define the contemporary United States of America as a feudal society on the basis of the feudal nature of higher education. For all of its display of feudal tradition,

US universities have become an extension of the corporate sector to a large extent (see Perelman 2002).

Just as a pre-capitalist society may contain elements that we consider to be characteristic of a capitalist society, a capitalist society typically contains remnants of pre-capitalist societies, as well as the seeds of a future communist society. In fact, no society has ever been thoroughly capitalistic, in the sense that all other forms of social relations have been expunged.

People unfamiliar with the arcane world of academia might have trouble in regarding higher education as feudal. Let me refer you to a more obvious example. In the United States of America until the middle of the nineteenth century, slaves and independent farmers performed the majority of work. Neither of these groups was involved in a typical capitalist social relation of working for a wage. The pre-capitalist nature of independent farming, and even more so of slavery, is probably easier to accept (see Chapter 12).

Let me return to the question of dominance. I would also argue that capitalist social relations were dominant in the mid-nineteenth century, despite the undeniable importance of slaves and independent farmers. Notwithstanding their numbers and their contribution to overall production, the independent farmers, as well as the slaves and the plantation owners, were becoming increasingly marginalised.

At that time, the social relations symbolised by the emerging Republican Party were increasingly defining the direction that the United States would follow. In effect, the social relations associated with independent farming and slavery were a part of the economy that was losing its vigour, while the more urban, industrial sectors were gaining the strength and confidence that allowed them to take the lead. So, dominance is not merely a matter of the number of people within a particular social relation.

When exactly did the industrial sector become more vigorous than the agrarian sector? Such a question is every bit as unanswerable as the earlier question about the dating of the origin of capitalism. Moreover, the nineteenth-century industrial sector could not have existed without its agrarian counterpart. The farms fed the workers in the factories and represented the major outlet for their goods.

Disputes about the origins of capitalism are even more contentious than those regarding the timing of the emergence of the capitalist. Again, social relations provide the key. Adam Smith, as well as a French economist, Turgot, proposed that capitalism began as a voluntary outgrowth of a natural tendency of people to engage in mutually beneficial trades. The theoretical approach of both Smith and Turgot floundered on one particular point: the emergence of the wage–labour relationship. Are we really to believe that some people voluntarily adopted the role of poor people who had nothing to trade except their own capacity to labour?

PRIMITIVE ACCUMULATION

Nobody put the contrary theory more forcefully than Karl Marx, who attributed the rise of capitalism to violent acts that expropriated the land and other property of the great mass of the population. The concept of primitive accumulation began in confusion and later settled into an unfortunate obscurity (see Perelman 2000). The seemingly Marxian expression, 'primitive accumulation', originally began with Adam Smith's assertion that 'the accumulation of stock must, in the nature of things, be previous to the division of labour' (1976, II.3, p. 277). Marx emphasised primitive accumulation, the expropriation of land and other means of production, rather than the accumulation of stock through saving and investment.

For Marx (1977, p. 926), 'capital comes dripping from head to toe, from every pore, with blood and dirt'. Workers were 'tortured by grotesquely terroristic laws into accepting the discipline necessary for the system of wage-labour' (p. 899). Where Smith scrupulously avoided any analysis of social relations, Marx produced an elaborate study of the connection between the development of capitalistic social relations and the so-called primitive accumulation.

Nonetheless, Marx played down primitive accumulation because it detracted from his more fundamental analysis of capitalism. With primitive accumulation, capitalists steal property from people. Their behaviour merits disapproval because they act in a way that seems to be unfair.

Marx wanted to show how the normal, legal functioning of the market, aside from any individual unfair behaviour, expropriates value from the working class. Marx did not attribute this capture of surplus value to 'bad' behaviour on the part of individuals, but to the impersonal functioning of a class system.

According to the typical reading of primitive accumulation, this original expropriation occurred in the distant pre-capitalist past. After the completion of this initial burst of primitive accumulation, a small group of people could function as capitalists.

By downplaying primitive accumulation, Marx pointed his readers to the importance of the way that the normal buying and selling of labour power robbed the working class. Unfortunately, that approach also obscured the way that primitive accumulation continued to occur alongside the capitalist system.

Most obviously, the colonial powers stole massive amounts of property throughout the colonial period. In fact, Marxist economists have long debated, without resolution, the extent to which this second wave of primitive accumulation rather than the ongoing accumulation of capital through the exploitation of wage labour was responsible for the massive development of the late eighteenth and early nineteenth centuries in the core capitalist economies.

Primitive accumulation was, in fact, much more a sophisticated process than a once-and-for-all event. Early capitalists realised that if people in the countryside were able to provide for their own needs, they would never work for wages. However, the more they were able to provide for themselves, the less money they would need in order to survive. So the early capitalists went to considerable lengths to create a situation in which people had enough means of production to allow wages to go as low as possible without giving them an option to survive without engaging in wage labour.

So, in effect, capitalists benefited from the lack of dependence on the market. Later, when they became more concerned about finding new outlets for their products, they benefited from people being more dependent on the market.

Today, when capital comes up against new limits to its ability to accumulate, once again it seeks to take over public resources through privatisation and other means. Schools, health care, water supply, even prisons are coming under the control of corporations. In addition, as families become more pressed to survive with one income, more and more women enter the labour force. In the process, many functions that the household would traditionally provide, such as food preparation and childcare, are being increasingly purchased as commodified services.

Smith, of course, had something other in mind than Marx's vision of violent primitive accumulation. He was attempting to speculate on the origins of capital. By showing how one individual could accumulate capital without impoverishing another, Smith could account for the voluntary origins of capitalism, which he so desperately wished to find. Sadly, he failed miserably in this endeavour.

Nonetheless, in his futile effort, Smith described the first act of capital accumulation as 'original accumulation'. By the time the term was translated from English into German and back into English again, original accumulation had become primitive accumulation, a more ominous-sounding expression, suggesting the brutality associated with the expropriation.

Without a doubt, primitive accumulation was an essential component of the rise of capitalism throughout Western Europe, especially England. Gold flowed from the mines of Latin America, black slaves from Africa, and wealth of all kinds from Asia, which, on the eve of capitalism, had an economy more technologically advanced than that of Europe. On top of the plunder from other nations, the English gentry confiscated the land from most of its own people, consigning them to work for wages in the factories and on the farms.

Notice that although the social relations between labour and capital define capitalism, primitive accumulation operates on the level of property relations. In that sense, primitive accumulation might seem to be unrelated to capitalism, even though it is a necessary precondition.

THE CONTINUING VIOLENCE OF CAPITALISM

Just as the act of procreation differs from the process of raising a child, the way primitive accumulation fostered the creation of capitalism seems to differ considerably from the way a functioning capitalist economy accumulates wealth. The analogy is admittedly imperfect because primitive accumulation continues to make a contribution to the accumulation of capitalist wealth, while the act of procreation occurs at a specific moment in time.

Nonetheless, the analogy is appropriate in the sense that the way a functioning capitalist economy typically accumulates wealth on the basis of the social relations between capital and labour differs significantly from the often openly violent process of primitive accumulation. Within the system of primitive accumulation, an exchange of equivalents is unnecessary. The more powerful party merely takes what it wants.

Within the mindset of ordinary people, primitive accumulation represents a violation of human rights and property rights. The response is often a sort of moral outrage. In contrast, under capitalism the law of value prevents arbitrary actions by individuals. Consequently, capitalist accumulation through the exploitation of labour gives the appearance of fairness. Workers ask for jobs. Employers give work in exchange for wages. Everyone supposedly gets something from the exchange.

Appearances can be deceptive. Workers use a considerable part of their wages to buy necessities. Were they slaves rather than wage labourers, their owners would have to supply them with food, clothing, and shelter if they were to continue to be able to perform productive work.

Few people would claim that slavery was fair because of an exchange of equivalents. Most people would say that the exchange between master and slave does not represent an equivalence.

Marx attempted to show with objective analysis that the exchange between labour and capital creates a similar type of exploitation, but he was intent on transcending the moral question of fairness. Like slaves, workers did what they did because of the objective condition in which they found themselves. Stripped of their capacity to provide for themselves, they had little choice but to work for others.

The violence of the capitalist system may be hidden, but it is violent nonetheless. This particular arrangement is not necessarily any more or any less fair than primitive accumulation. Its main fault is that it prevents both people and society from achieving their potentials.

For Marx, what was important was the ability to analyse the process in order to create something superior. He believed that capitalism would fail not because it was unfair, but because its short-comings would become so obvious that people would throw it aside for a more rational organisation of society. Moral outrage about fairness was beside the point.

For that reason, Marx placed the material on primitive accumulation at the end of his book (Marx 1977, part 8). It appeared to be less of a conclusion than an appendix, outside of the main thrust of the book, except for one brief section that gave a beautiful analysis of the relationship between primitive accumulation and ongoing capitalist accumulation.

There, Marx described the work of Edward Gibbon Wakefield, an Englishman who spent some time in jail for attempting to abduct and marry a young schoolgirl (Marx 1977, ch. 33). While languishing in prison, Wakefield learned about Australia because many of his fellow inmates were to be transported there. Wakefield found that many of the prisoners who were expected to do labour once they got to Australia instead went inland and began to work land on their own. He realised that an abundance of free resources was incompatible with getting workers to labour for others.

From prison, Wakefield wrote a book purporting to be letters from an Australian (Wakefield 1829). He convincingly argued that the creation of a viable capitalist economy in Australia would require putting much of the land off-limits, until the population reached the point that free land would no longer be available. The book was so successful that Wakefield became a major figure in the organisation of British colonisation. Under his leadership, New Zealand followed his recommendations, preventing a rapid mass migration of people into the inland regions of the country.

For Marx, Wakefield's story was a parable of the capitalist economy. What we now call free markets based on wage labour are only possible when the preconditions make labour unfree. Nonetheless, aside from the Wakefield section, most of Marx's materials on primitive accumulation seem somewhat unconnected with his more theoretical analysis of the nature of a capitalist economy.

This separation of the historical creation of capitalism and the ongoing functioning of capitalism had some negative consequences. It suggested that the beginning of capitalism was an event (like procreation), after which primitive accumulation ceased. This 'Big Bang' never happened. Instead, primitive accumulation initiated a long process.

Recall that that which defines capitalism is a particular type of social relation. This social relation goes beyond merely placing the worker in the employ of another; it requires that this arrangement be seen as in some sense 'normal'.

The artificial separation of the stages of capitalist development is unfortunate in another respect. While primitive accumulation was a necessary step in the initial creation of capitalism, it actually continues to this day. For example, at the time of this writing, petroleum and mining companies are displacing indigenous people in Asia, Africa, Latin America and even in the United States. Emphasising the social relations of advanced capitalist production to the exclusion of the ongoing process of primitive accumulation obscures the fact that the struggles of the Ogoni people in Nigeria or the Uwa in Colombia are part of the same struggle as that of exploited workers in Detroit or Manchester.

THE CONTINUING ROLE OF PRIMITIVE ACCUMULATION

The artificial separation of capitalism into stages makes another fundamental element murkier than it need be: even today, access to resources depends, more often than not, on primitive accumulation. In this sense, capitalism has not changed a great deal over the centuries.

While capitalist production may seem to reside within the confines of the world of advanced technology, run by advanced electronic controls, this system ultimately depends on access to cheap raw materials – especially petroleum. For example, even the supposedly weightless economy still depends on colossal server farms. One, being constructed near Seattle, covers about 13 acres (McKay 2000).

These server farms have power concentrations of 100 watts per square foot. Ten square feet consume enough power to supply a typical home. These data centres are expected to cover an estimated 50 million square feet by 2005, but even so their demand will amount to slightly more than 1 per cent of the US electricity demand by that time (Bryce 2000).

The material demands of the New Economy go well beyond fossil fuels. The United States used about 1 billion tonnes of materials in 1990, such as iron, copper, sulphur and phosphorus, and hydrocarbon fossil fuels, as well as other materials that are mined and used in the production of goods, but excluding some (crushed) stone that is used to build roads and other structures (Wernick et al. 1996). Access to these resources depends, more often than not, on primitive accumulation. In this sense, capitalism has not changed a great deal over the centuries.

REFERENCES AND FURTHER READING

Battalio, R.C., Kagel, J.H and McDonald, D.N. (1985) 'Animal's Choices over Uncertain Outcomes', *American Economic Review* 75(4), pp. 597–613.

Bryce, R. (2000) 'Power Struggle' *Interactive Week*, 18 December (www4.zdnet.com:80/intweek/stories/news/0,4164,2665339,00.html).

Marx, K. (1977) *Capital*, vol. 1. New York: Vintage.

McKay, J. (2000) 'Server Farms Strain Local Grids: Jurisdictions Are Facing Huge Power Demands From These Digital Warehouses' *Government Technology News*, 29 September (www.govtech.net/news/features/feature_sept_29.phtml).

Perelman, M. (2000) *The Invention of Capitalism: The Secret History of Primitive Accumulation*. Durham: Duke University Press.

Perelman, M. (2002) *Steal This Idea: The Corporate Capture of Creativity*. New York: Palgrave.

Smith, A. (1776) *The Nature and Causes of the Wealth of Nations*. Oxford: Oxford University Press, 1976.

Wakefield, E.G. (1829) 'Letter from Sydney: The Principle Town of Australasia', in *The Collected Works of Edward Gibbon Wakefield*, ed. M.F. Lloyd Pritchard. Glasgow and London: Collins, 1968.

Wernick, I.K., Herman, R., Govind, S. and Ausubel, J.H. (1996) 'Materialization and Dematerialization: Measures and Trends', *Daedalus* 125(3), pp. 171–98 (phe.rockefeller.edu/Daedalus/Demat).

9 Globalisation and the State: Where is the Power of Capital?

Ellen Meiksins Wood

Anti-capitalist movements, from the earliest days of socialism to anti-globalisation protests today, have always encountered one fundamental problem: power in capitalist societies is so diffuse that it is difficult to identify a target for opposition. Of course, workers in any individual enterprise can fight against their employer for better terms and conditions of work. Sometimes many enterprises in a single industry can be challenged by their common trade union. But it is much harder to locate a point in capitalist society where power is concentrated in such a way that resistance and opposition can be effectively directed against class domination by capitalists in general, or against the logic of the capitalist system, which puts 'profits before people'.

THE ELUSIVE POWER OF CAPITAL

In noncapitalist class societies, it is not usually so difficult to identify the locus of power. Find the source of military and political coercion and you will generally find economic power too. Capitalism is distinctive among class societies in this respect. Capitalists – unlike, say, feudal lords – generally need no direct control of coercive military or political force to exploit their workers who, because they are propertyless and have no direct access to the means of production, must sell their labour power in exchange for a wage in order to work and to live (see Chapter 1).

To be sure, capitalists ultimately depend on coercion by the state to underpin their economic powers and their hold on property, to maintain social order and conditions favourable to accumulation. But there is a more or less clear division of labour between the exploitative powers of the capitalist and the coercive powers of the state. In capitalist societies, it is even possible to have universal suffrage without fundamentally endangering capitalist economic power, because that power does not require a monopoly on political rights.

There is even a sense in which only capitalism has a distinct 'economic' sphere at all. This is so both because economic power is separate from political or military force and because it is only in capitalism that 'the market' has a force of its own, which imposes on everyone, capitalists as well as workers, certain impersonal systemic requirements of competition and profit maximisation. Because all economic actors depend on the market for everything they need, they must meet its requirements in order to survive, irrespective of their own personal needs and wants.

This brings us to the second reason for the diffusion of power in capitalism. Coercion in capitalist societies is exercised not only personally and directly by means of superior force but also indirectly and impersonally by the compulsions of the market. The dominant class, with the help of the state, can and does certainly manipulate those compulsions to its own advantage, but it is impossible to trace them to a single source of power.

While capital does require support by state coercion, the power of the state seems to be circumscribed by the powers of capital. A very wide range of social functions – not only the organisation of production, but the distribution of resources, the disposition of labour and the organisation of time itself – is removed from political or communal control, and placed in the economic sphere, either under the direct control of capital or subject to the impersonal 'laws' of the market.

In fact, one of the most important characteristics of capitalism is that the economic hegemony of capital can extend far beyond the limits of direct political domination. This is true not only of class relations between capital and labour but also of relations between imperial and subordinate states. We have already noted capital's ability to dominate labour by purely 'economic' means and without direct political rule or judicial privilege. This contrasts sharply with noncapitalist classes which depended on 'extra-economic' powers of coercion. Such classes relied on their superior coercive force, on their political and military power and privilege, to extract surplus labour, typically from peasants who, unlike capitalist wage labourers, remained in possession of the means of production, either as owners or as tenants. There is an analogous difference between noncapitalist and capitalist imperialisms. Old colonial empires dominated territory and subject peoples by means of 'extra-economic' coercion, by military conquest and often direct political rule. Capitalist imperialism can exercise its rule by economic means, by manipulating the 'laws' of the market, including the weapon of debt (see Chapters 8 and 13).

The state remains vital to this kind of domination, in ways that will be discussed in what follows. But the separation between economic and political domination clearly presents problems for oppositional struggles. All of this has inevitably affected the nature of opposition and class struggle. It is, for instance, no accident that modern revolutions have occurred not in advanced capitalist societies but in societies where the state has presented a visible target, with a prominent role in direct exploitation. But as capitalism develops into its mature industrial form, there tends to be a growing concentration of class struggle in the workplace and a growing separation between 'industrial' and 'political' struggles.

This distinctive relation between economic and political spheres has *always* posed a problem for anti-capitalist movements. But as long as there was some more or less clear connection between national economies and national states, there remained a clear possibility of challenging the power of capital not only in the workplace but at a point of concentration in the state. At the very least, pressure could be put on the state by organised oppositional forces, most particularly the labour movement, to undertake policies that would to some extent ameliorate the worst effects of capitalism. The division of labour between political and economic spheres could even work to the advantage of subordinate classes, and the balance of class forces within the state itself might shift significantly in favour of the working class, so that, even while the state remained within the constraints of the capitalist system, it could act more positively in the interests of workers. There was even a hope that seizure of state power would make possible a more complete social transformation, the replacement of capitalism by socialism.

But today, it seems that even the most limited of these possibilities hardly exist. At first glance, the separation of economic from political power seems an even greater, and perhaps insurmountable, problem in today's 'globalised' economy than ever before. Transnational capital seems to have escaped the boundaries of the nation state, the power of capital seems to have become even more diffuse, and the problem of locating and challenging the centre of capitalist power has apparently become harder than ever. It seems to be everywhere and nowhere.

Many of today's anti-capitalist protestors have therefore sought their principal targets in transnational organisations. Some of the most well-known critics of globalisation, at least in the dominant capitalist economies, characterise it mainly as a development driven and dominated by transnational corporations, whose infamous brand names – Nike, McDonald's, Microsoft, and so on – are the symbols of today's global capitalism. At the same time, these critics seem to assume that the services traditionally performed by the nation state for national capital must now be performed for transnational corporations by some kind of global state. In the absence of such a state, the political work of global capital is apparently being done by transnational institutions such as the WTO, the IMF, or G8 summits. Anti-capitalist movements acting on these assumptions have targeted transnational corporations by such means as consumer boycotts, sabotage and demonstrations. But above all, they have directed their oppositional energies against supranational organisations which appear to be the institutions that come closest to representing the political arm of global capital, in the way that the nation state has traditionally represented national capital.

These 'anti-capitalist' movements have been effective in bringing to light the devastating effects of 'globalisation'. They have raised the consciousness of many people throughout the world, and they have offered the promise of new oppositional forces. But it may be that in some respects they are based on faulty premises.

GLOBALISATION: MORE GLOBAL OR MORE CAPITALIST?

The conviction that global corporations are the ultimate source of globalisation's evils, and that the power of global capital is politically represented above all in supranational institutions like the WTO, may be based, first, on the assumption that global capitalism behaves the way it does because it is global, rather than (or more than) because it is capitalist. The principal task for oppositional forces, then, is to target the instruments of capital's global reach rather than to challenge the capitalist system itself.

In fact, many participants in movements of this kind are not so much anti-capitalist as anti-'globalisation', or perhaps anti-neoliberal, or even just opposed to particularly malignant corporations. They assume that the detrimental effects of the capitalist system can be eliminated by taming global corporations or by making them more 'ethical', 'responsible', and socially conscious.

But even those who are more inclined to oppose the capitalist system itself may assume that the more global the capitalist economy becomes, the more global the political organisation of capital will be. So, if globalisation has made the national state increasingly irrelevant, anti-capitalist struggles must move immediately beyond the nation state, to the global institutions where the power of global capital truly lies.

We need to examine these assumptions critically, not because anti-capitalist movements are wrong in their conviction that transnational corporations are doing great damage and need to be challenged, or that the WTO and the IMF are doing the work of global capital – which is certainly true. Nor are these movements wrong in their internationalism or their insistence on solidarity among oppositional forces throughout the world. We need to scrutinise the relation between global capital and national states because even the effectiveness of international solidarity depends on an accurate assessment of the forces available to capital and those accessible to opposition.

Let us start from the premise that global capitalism is what it is not only because it is global but, above all, because it is capitalist. The evils we associate with globalisation – the social injustices, the growing gap between rich and poor, 'democratic deficits', ecological degradation, and so on – are there not simply because the economy is 'global', or because global corporations are uniquely vicious, or even because they are exceptionally powerful. These problems exist because capitalism, whether national or global, is driven by certain systemic imperatives, the imperatives of competition, profit maximisation and accumulation, which inevitably require putting 'exchange value' above 'use value' and profit above people. Even the most 'benign' or 'responsible' corporation cannot escape these compulsions but must follow the 'laws' of the market in order to survive – which inevitably means putting profit above all other considerations, with all its wasteful and destructive consequences. These compulsions also require capital's constant self-expansion (see Chapters 1 and 4). Globalisation itself, however much it has intensified these imperatives, is their result rather than their cause.

These systemic imperatives can certainly operate through the medium of specific transnational corporations, but, as one commentator has put it, 'corporations, as powerful as they are, are only vehicles for capitalists ... It's often assumed that corporations are a power in themselves, rather than a particular way in which capitalists organise their wealth.'[1] Any particular organisation of capitalist wealth, such as Monsanto, can be challenged, even wrecked, but the capitalists involved can simply restructure their wealth, restore their profits in another form, and resume their destructive activities – all of which Monsanto did, very soon after what was one of the most effective anti-globalisation campaigns.

This is not to say that such campaigns are fruitless. They should certainly be supported. The point is simply to recognise that targeting even the most destructive corporations still is very far from challenging the sources of capitalist power or capitalism's systemic compulsions, the imperatives that compel capitalists (whether they are evil or relatively benign and responsible) to accumulate relentlessly and constantly maximise their profits. Targeting, or even wrecking, specific corporations or supranational institutions like the WTO certainly has the advantage of complicating the daily life of capital, but it brings us not much closer to the core of capitalism.

A DECLINING NATION STATE?

If we start from the premise that the problem is not this or that corporation, nor this or that international agency, but the capitalist system itself, we are, of course, left with the problem of tracing capitalist imperatives to a source that is accessible to challenge. How do you fight a whole system? No one can deny that this remains an intractable problem. But at the very least, we can raise questions about whether the global scope of capital has put it so far beyond the reach of the national state that the nation state is no longer a major source of capitalist power, a major target of resistance nor a potential instrument of opposition.

We can consider, first, the main functions traditionally performed by the nation state for domestic capital and ask whether these functions have been transferred to transnational organisations.

In every class society, where one class appropriates the surplus labour of another, there are two related but distinct 'moments' of class exploitation: the appropriation of surplus labour and the coercive power that enforces it. In noncapitalist societies, these tended to be more or less united. The separation of 'economic' and 'political' spheres in capitalism has meant that these two moments have been effectively divided between private enterprises and the public power of the state or even between state enterprises acting on capitalist principles and the coercive arm of the state. To be sure, any capitalist enterprise has at its disposal an array of disciplinary mechanisms, as well as internal organisational hierarchies, to keep workers in line and at work; and the most effective sanction available to capital is its ability to deny the worker access to the means of labour, that is, its ability to deny the worker a job and a wage, to dismiss workers or to close enterprises altogether. But the ultimate sanction that sustains the system as a whole belongs to the state, which commands the legal authority, the police and the military power to exert direct coercive force. While capitalists have used their property to exploit propertyless workers in the 'economic' sphere, the state has maintained social order.

From the beginning, intervention by the state has been needed to create and maintain not only the system of property but also the system of propertylessness. State power has, of course, been needed to support the process of expropriation and to protect the exclusiveness of capitalist property. But the state has also been needed to ensure that, once expropriated, those without property in the means of production are available, when required, as labour for capital.

Here, a delicate balance has had to be struck. On the one hand, the state must help to keep alive a propertyless population which has no other means of survival when work is unavailable, maintaining a 'reserve army' of workers through the inevitable cyclical declines in the demand for labour. On the other hand, the state must ensure that escape routes are closed and that means of survival other than wage labour for capital are not so readily available as to liberate the propertyless from the need to sell their labour power when they are needed by capital.

This balancing act has been a major function of the state since the earliest days of capitalism. The state has also performed another major and related function: controlling the mobility of labour, while preserving capital's freedom of movement. Although the movement of labour across national boundaries has been severely restricted, controlling labour's mobility need not mean keeping workers immobile. It may mean getting them to move where capital most needs them. Sometimes, especially in the early days of industrial development, the state has helped to uproot labour, to separate it from local attachments. But even when the state has made labour freely available by movements within and across borders if required, such movements have always been rigorously controlled. It has been one of the state's most essential functions to keep a firm grip on the mobility of labour, so that the movements of labour enhance, rather than endanger, capitalist profit.

Capitalism is, by its nature, an anarchic system, in which the 'laws' of the market constantly threaten to disrupt the social order. Yet capitalism needs stability and predictability in its social arrangements probably more than any other social form. The nation state has from the beginning provided that stability and that predictability by supplying an elaborate legal and institutional framework, backed up by coercive force, to sustain the property relations of capitalism, its complex contractual apparatus and its intricate financial transactions.

The question then is whether 'global' capitalism has found other and better means than the nation state to perform all, or indeed any, of these basic functions. Even a moment's reflection should make it clear that no other institution, no transnational agency, has even begun to replace the nation state as a coercive guarantor of social order, property relations, stability or contractual predictability, or any of the other basic conditions required by capital in its everyday life. The state still provides the indispensable conditions of accumulation for global capital, no less than for very local enterprises; and it is, in the final analysis, the state that has created the conditions enabling global capital to survive and to navigate the world. It would not be too much to say that the state is the *only* non-economic institution truly indispensable to capital. While we can imagine capital continuing its daily operations with barely a hiccup if the WTO were destroyed, it is inconceivable that those operations would long survive the destruction of the local state.

For all the attacks on the welfare state launched by successive neoliberal governments throughout our era of 'globalisation', it cannot even be argued that global capital has been able to dispense with the social functions performed by nation states since the early days of capitalism. Even while labour movements and forces of the left have been in retreat, with so-called social democratic governments joining in the neoliberal assault, at least a minimal 'safety net' of social provision has proved to be an essential condition of economic success and social stability in advanced capitalist countries. At the same time, developing countries that may have been able to rely more on traditional supports, such as extended families and village communities, have been under pressure to shift at least some of these functions to the state, as the process of 'development' and the commodification of life have destroyed or weakened old social networks although this, in turn, has made them more vulnerable to privatisation, under pressure from the agencies of global capital.

Oppositional movements must struggle constantly to maintain anything close to decent social provision. But it is hard to see how any capitalist economy can long survive, let alone prosper, without a state that to some extent, however inadequately, balances the economic and social disruptions caused by the capitalist market and class exploitation. Globalisation, which has further undermined traditional communities and social networks, has, if anything, made this state function more, rather than less, necessary to the preservation of the capitalist system. This does not mean that capital will ever willingly encourage social provision. It simply means that capitalism's hostility to social programmes, as necessarily a drag on capital accumulation, is one of its many insoluble contradictions.

GLOBAL CAPITAL, GLOBAL STATE?

Are there, nonetheless, certain new or growing functions that are specifically global in their scope, which must be administered by transnational agencies instead of by nation states? No one can doubt that movements of capital across national boundaries are frequent and breathtakingly rapid in today's global economy, or that new supranational institutions have emerged to facilitate those movements. But whether that means that markets are substantially more globally integrated than ever before, or, even if they are, that the role of the nation state has diminished accordingly, is another question.

The first and most elementary point is that so-called 'transnational' corporations generally have a base, together with dominant shareholders and boards, in single nation states and depend on them in many fundamental ways. Beyond that simple point, some commentators have argued that, according to various measures of integration, globalisation is far from advanced, and in important respects less so than in previous eras – for instance, in the magnitude of international trade as a share of gross domestic product, or global exports as a proportion of the global product.

But let us accept that the speed and extent of capital movements, especially those that depend on new information and communication technologies, have created something new. There remains one overriding indication that the global market is still far from integrated: the fact that wages, prices and conditions of labour are still so widely diverse throughout the world. In a truly integrated market, market imperatives would impose themselves universally, to compel all competitors to approximate some common social average of labour productivity and costs, in order to survive in conditions of price competition.

This apparent failure of global integration is not, however, a failure of globalisation so much as a symptom of it. In fact, globalisation has been as much about *preventing* as promoting integrated markets. The global movements of capital require not only free transborder access to labour, resources and markets but also a kind of economic and social fragmentation that enhances profitability. And here again, it is the nation state that must perform the delicate balancing act between opening borders to global capital and deterring a kind and degree of integration that might go too far in levelling social conditions among workers throughout the world.

It cannot even be said that global capital would gain most from levelling the costs of labour downward by subjecting workers in advanced capitalist countries to the competition of low-cost labour regimes. This is certainly true, up to a point. But, apart from the dangers of social upheaval at home, there is the inevitable contradiction between capital's constant need to drive down the costs of labour and its constant need to expand consumption, which requires that people have money to spend. This, too, is one of the insoluble contradictions of capitalism. But, on balance, global capital benefits from uneven development; and the fragmentation of the world into separate economies, each with its own social regime and labour conditions, presided over by more or less sovereign nation states, is no less essential to 'globalisation' than is the free movement of capital. Not the least important function of the nation state in globalisation is to enforce the principle of nationality that makes it possible to manage the movements of labour by means of strict border controls and stringent immigration policies, in the interests of capital.

But the first and most basic condition of globalisation is imposing market imperatives throughout the globe. This does not mean that imperial powers will encourage the development of capitalist economies like their own everywhere. It simply means that subordinate economies must be made vulnerable to the dictates of the capitalist market, by means of certain social transformations – such as, for example, the transformation of peasants into market-dependent farmers, as subsistence agriculture is replaced by specialisation in single cash crops (while, of course, the metropolitan powers protect their own domestic agriculture by huge subsidies and import controls). Bringing about such social transformations has been a major function of capitalist imperialism since its inception, and the indispensable instrument has been the nation state.

Older forms of imperialism, as we have seen, depended directly on conquest and colonial rule. Capitalism has extended the reach of imperial domination far beyond the capacities of direct political rule or colonialism, simply by imposing and manipulating the operations of a capitalist market. Just as capitalist classes need no direct political command over propertyless workers, capitalist empires can rely on economic pressures to exploit subordinate societies. But just as workers had to be made dependent on capital and kept that way, so subordinate economies must be made and kept vulnerable to economic manipulation by capital and the capitalist market – and this can be a very violent process.

The most recent methods of imposing market imperatives are familiar in countries that have undergone 'structural adjustment'. But, in various forms, the process goes back to the earliest days of capitalist imperialism. England, even in the late sixteenth century, was already experimenting with this imperialist strategy, notably in Ireland. And from the beginning, capitalist imperialism has been affected by one of the main contradictions of capitalism: the need to impose its economic 'laws' as universally as possible, and, at the same time, the need to limit the damaging consequences that this universalisation has for capital itself. Capitalism is driven by competition, yet capital must always seek to thwart competition. It must constantly expand its markets and constantly seek profit in new places, yet it typically subverts the expansion of markets by blocking the development of potential competitors (as it did in Ireland, already in the seventeenth century).

The nation state has been an indispensable instrument in the process of spreading capitalist imperatives, not only in the sense that the military power of European nation states has carried the dominating force of capital to every corner of the world, but also in the sense that nation states have been the conduits of capitalism at the receiving end too. Indeed, for all the globalising tendencies of capitalism, the world has become more, not less, a world of nation states, not only as a result of national liberation struggles but also under pressure from the imperial powers. These powers have found the nation state to be the most reliable guarantor of the conditions necessary for accumulation, and the only means by which capital can freely expand beyond the boundaries of direct political domination. As market imperatives have become a means of manipulating local elites, local states have proved to be far more useful transmission belts for capitalist imperatives than were the old colonial agents and settlers who originally carried the capitalist market throughout the world.

But this mode of imperialism also reveals a curious contradiction at the heart of capitalism, especially in today's 'globalised' form. Capitalism has a unique drive for self-expansion. Capital cannot survive without constant accumulation, and its requirements relentlessly drive it to expand its geographic scope too. From its earliest days, capital, while always needing the support of nation states, has been driven beyond national borders. The separation of the 'economic' and the 'political' has made it possible for the economic reach of capital to extend much further than its political grasp – in a way that was never possible for earlier forms of economic exploitation which depended directly on military power and political rule.

Globalisation is taking this separation, this division of labour, between the economic and the political to its utmost limit. But it is *not* making the nation state less relevant to capital. Despite the emergence of various transnational institutions – which are, after all, little more than instruments of certain nation states, and one in particular – there is little evidence that global capital is losing its dependence on national states. It seems very unlikely that there will ever be a capitalist state that corresponds to the global economy.

Globalisation, then, does not mean the decline of the nation state. If anything, the new form of imperialism we call globalisation is more than ever an imperialism that depends on a system of multiple states. Precisely because the imperialism of globalisation depends on extending purely economic hegemony and market imperatives far beyond the reach of any single state, it is especially dependent on a plurality of subordinate states to enforce those imperatives and to create the climate of legal and political order, the stability and predictability, that capital needs in its daily transactions.

GLOBALISATION AND WAR

The US attack on Afghanistan is a dramatic illustration of the relation between globalisation, imperialism and the nation state. The most obvious point is that the state is revealing its ultimate power, the power to make war. It is at the same time revealing other powers which we have been told no longer exist. Having insisted that the movements of capital have escaped state control, for instance, imperial states are now freely proposing to freeze the assets of terrorist groups. But the war reveals other more essential connections between globalisation and the state.

The history of the whole region is, to begin with, testimony to the imperialist practice of creating and manipulating states to serve the purposes of capital. Osama bin Laden himself is in many ways a product of that practice. He was formed in Saudi Arabia, and the Saudi Arabian state is his principal target. That state developed into its present form after the carve-up of the Middle East, engineered especially by Britain and France as the Ottoman empire disintegrated. Since then, Western states, and increasingly the United States of America, have propped up this repressive regime, and others, mainly to ensure the oil supply that capital so desperately needs. The United States also, of course, supported religious extremists, such as those who created the Taliban, and bin Laden himself, in their efforts to ensure a friendly, anti-Soviet regime in Afghanistan. This imperial practice of state formation has certainly not been displaced by globalisation. Some commentators have suggested, for instance, that the United States is now exploiting the opportunity to reshape Afghanistan, with an eye to the huge oil and gas reserves of Central Asia (though it must be said that the US has shown little interest in the process of 'nation building' there, as it leaves its 'allies' and the UN to clean up its mess).

But there is more to the war than such specific imperial objectives. This war, like others in recent years, has a more general objective. The military power of the USA, by far the most powerful coercive force the world has ever known and the closest thing to a global state, is certainly the ultimate enforcer of globalisation. Yet try to imagine high-tech bombs, however 'smart', acting as the day-to-day regulator of a complex legal and contractual order, enabling the property transactions and financial dealings that are capitalism's daily bread, to say nothing of the everyday relations between capital and labour. For that, local states are indispensable.

But the trouble with a system of multiple states is that it creates great potential for disorder, resistance and opposition. Those states are subject to their own internal pressures and oppositional forces, and no single military power, not even the United States, with or without its allies, can ensure the compliance of so many states. Not even the most advanced military force can keep this global system in line all at once, by means of constant direct coercion.

Controlling a whole global economy, all the time and everywhere, and the many states that are needed to keep it working, is a very different matter from the old imperialist task of capturing territory or dominating particular states, with finite boundaries. So one way of keeping states in line is regularly to display the military power of the United States and demonstrate that, if it cannot be everywhere all the time, it can go anywhere at any time and do great damage.

The need to shape the world's political environment – both in subordinate states and in other advanced capitalist states which are implicated in military alliances[2] – may help to explain why the United States has so often embarked on military actions with no clear goal or exit strategy, and in situations not susceptible to military solutions. It may also help to explain the US practice of waging war without risking the lives of its own forces, even when massive air attacks cannot achieve the professed objectives.

The attack on Afghanistan will certainly not end terrorism. It is far more likely to increase terrorist violence. For that matter, even installing a friendly and stable regime in Afghanistan (assuming that the United States cares) is very far from guaranteed. But military action without apparent purpose may be an end in itself. Warmongers in Washington have been talking openly about the 'demonstration effect' of the war against Afghanistan, making it clear that, if they have little interest in what happens to Afghanistan, they have great hopes for the war's psychological effects on more important states, such as Iraq. It is especially revealing that hawks in the White House reportedly have a plan called 'Operation Infinite War', which calls for war without constraints of time or geography, and that US Vice President Cheney has warned that the 'war against terrorism' may last beyond our lifetimes.[3] Open-ended war against an invisible enemy is just what this new form of empire needs. The borderless empire of globalisation needs infinite war, war without boundaries, war that is *endless* in both purpose and time.

And yet, this kind of military strategy also exposes the growing contradiction between the global economy and the local political forces on which it depends. It reveals how dependent global capital is on local states and how unstable that structure of multiple states can be. At the very least, it confirms that the state is more than ever the point of concentration of capitalist power, and that therefore the state must more than ever be a target of oppositional struggles. Anti-capitalist demonstrations at meetings of the WTO or the IMF have been enormously important in many ways, but they are no substitute for the kind of *political* organisation that can truly challenge state power, and the balance of class forces the state represents, from both outside and inside the state.

REFERENCES AND FURTHER READING

Gowan, P. (1999) *The Global Gamble: Washington's Faustian Bid for World Dominance.* London: Verso.

Greenfield, G. (2001) 'Devastating, with a Difference: From Anti-Corporate Populism to Anti-Capitalist Alternatives', *Against the Current* 93, pp. 12–14.

NOTES

1. Greenfield (2001, pp. 13–14).
2. For a discussion of US efforts to manipulate its allies in this way, see Gowan (1999).
3. On 30 September, the *Observer* in London carried a special report by Ed Vulliamy, 'Inside the Pentagon'. Here are some of the highlights: 'As war begins in Afghanistan, so does the assault on the White House – to win the ear and signed orders of the military's Commander in Chief, President George W. Bush, for what Pentagon hawks call "Operation Infinite War" … The Observer has learnt that two detailed proposals for warfare without limit were presented to the President this week by his Defence Secretary Donald Rumsfeld, both of which were temporarily put aside but remain on hold. … They were drawn up by his deputy, Paul Wolfowitz … [T]he plans argue for open-ended war without constraint either of time or geography … [T]he Pentagon militants prefer to speak of "revolving alliances", which look like a Venn diagram, with an overlapping centre and only certain countries coming within the US orbit for different sectors and periods of an unending war. The only countries in the middle of the diagrammatic rose, where all the circles overlap, are the US, Britain and Turkey … Officials say that in a war without precedent, the rules have to be made up as it develops, and that the so-called "Powell Doctrine" arguing that there should be no military intervention without "clear and achievable" political goals is "irrelevant".'

10 Financial and Industrial Capital: A New Class Coalition

Suzanne de Brunhoff

We need to understand better how today's capitalism works, and how harmful it is to ordinary people across the world. This task requires a careful scrutiny of the social forces that might be able to oppose capitalism and bring about an alternative system.

International finance and money seem to be in a dominant position since the 1980s. All over the world, neoliberal policies have generated huge flows of short-term investment, currency speculation, financial instability and crises. The 'creation of shareholder value' has put economies under great strain. However between 1995 and 2000, there was growth and technical progress (and near full employment) in the United States, which looked like the model of a 'new economy' (see Chapter 16). We have to examine the relationship between capitalist production and finance, and the class divisions that are at stake today.

THE 'TRIUMPH OF FINANCIAL CAPITAL'?

Many opponents of capitalist globalisation denounce the vices of financial hegemony. Contrary to the accumulation of productive, 'real' capital, financial activity transforms private wealth into highly profitable investments, but it needs liquidity and mobility. This entails the 'short-termism' of speculators and the parasitism of rentiers, which are counterposed to the productive contribution of economic agents, entrepreneurs and workers. The 'activities of a casino' are harmful to industrial development and growth.

A key feature of this financial hegemony is the overpowering demand of shareholders for high returns and high short-term company profits. Bank loans and bonds are now more dependent on international short-term profits. They provide credit to enterprises or countries that are able to pay high private returns. When these returns are declining, big international banks cut credit lines, and money flows back to safe havens. Then weak currencies are devalued, which entails financial crises and economic recession in the 'emerging' countries.

Neoliberal policies are accused of paving the way to this financial hegemony. When did they start? At first they were political answers to the capitalist crisis of the 1970s, combining inflation and stagnation, and the depreciation of the dollar. The Bretton Woods system collapsed, and financial liberalisation started with the free floating of the main currencies (the dollar, the D-mark and the yen) and the end of exchange controls. Private financial markets now settled the rates of exchange.

The main political turn came in 1979–80, with Reaganomics in the United States and Thatcherism in the United Kingdom, and dis-inflation policies in Western Europe. The value of money was restored. These policies worked in favour of money capital and financial asset owners. However, we need to understand the rela-tionship between industrial capital and real growth, and whether economic stagnation was the price to pay for 'the triumph of financial capital' (Sweezy 1994, p. 2) since the 1980s.

A NEW COALITION BETWEEN FINANCIAL AND INDUSTRIAL CAPITALISTS

The visible domination of financial markets does not imply that industrial capital has lost its fundamental importance. Rather, we could say that a new capitalist coalition has emerged, in order to restore profitability after the crisis of the1970s. When shareholders required very high returns on their financial assets, bosses responded by reorganising production processes: downsizing, subcontracting, relocation of plants and so on. Since the 1980s, company restruc-turing has maximised profits *while treating employment and wages as adjustment variables*.

Everywhere labour was broken by the new profitability regime. In response to the profit crisis of the 1970s, the 'new economy' was driven by an international capitalist coalition of financial and industrial capitalists, with the help of neoliberal policies. Techno-logical progress blossomed. This complex process restored high profits and supported the emergence of a financial boom. Until 2000, it seemed that the growth of the 'new economy' in developed capitalist countries would last for ever.

The distribution of profits between industrial and financial capit-alists is often problematic, and conflicts of interest often arise. Each of these spheres of capital has its own form of expansion, instability and crisis. However, the accumulation of capital needs both of them. According to Marx (1867, p. 626), the 'credit system', including bank loans and financial markets, is 'a terrible weapon in the battle of competition, and is finally transformed into an enormous social mechanism for the centralisation of capital'. It operates by 'the violent method of annexation' of dispersed enterprises by preponderant centres of annexation, or by 'the smoother process of organising joint stock companies' (p. 627).

Great waves of mergers and acquisitions took place in the 1990s. They promoted qualitative changes in industrial production through the restructuring of plant and labour by subcontracting, outsourcing and relocation, within both developed and developing countries. During the same period, a huge centralisation of money was undertaken by mutual and pension funds. These quantitative and qualitative changes in the accumulation of industrial capital involve the active participation of finance.

This does not mean that finance capital is purely functional, or that financial markets act 'rationally'. Orthodox arguments in favour of free financial markets are well known. They claim that these markets are 'efficient' when asset prices incorporate immediately all available information, and that they rationally allocate savings to economic investment. Growth needs these markets. Without foreign investment, less developed countries could not grow, and industrial countries could not promote technical progress. Therefore, financial instability, currency crises and windfall speculative profits are part of the price to pay for global development and world wealth. However, there are large differences of opinion among orthodox analysts about the evaluation of financial assets.

There is a disparity between the market price of shares and the 'real' capital evaluation of companies, which is shown by different indices. One of them is the ratio of stockmarket evaluation of companies to the replacement cost of 'tangible assets', including plant, machines, and so on (Tobin's Q). In the 1990s, this ratio became very high, as share prices climbed. Does this mean that share prices were overvalued, when they were compared with the prices of the means of production? If this is the case, invoking 'irrational exuberance' does not explain how financial capital works.

There are different explanations of the 'irrationality' of the financial markets, according to different theories of economic value. The first belongs to the prevailing neoclassical theory of market equilibrium. This theory is unable to explain how financial markets work, because it does not understand the meaning of the demand for money and finance. 'Rational individuals are interested in the commodities they can produce and exchange. Their motives are measured in "real" terms (expressed in quantities of goods) not in "nominal" terms (values expressed in money)' (Arrow 1981, p. 139). The consequence is that money and the prices of financial assets cannot be determined within this theory's equilibrium analysis. Far from being 'irrational', they are simply not included in the economic rationale defined by this theory.

This failure has not prevented the construction of several mathematical models and measures of rational capital-asset pricing. All of them consider financial assets to be specific commodities bought and sold in a free market where rational individuals trade. The mathematical evaluation of the risks of losses and the chances of gains has become more and more developed. However, these theories remain within the financial sphere; therefore, they cannot explain how the prices of financial assets are related to economic 'fundamentals', if at all.

J.M. Keynes (1936) presented a different analysis of the irrational behaviour of financial markets. He introduced the theory of the demand for money and liquidity, and of different groups of economic agents. The irrationality of financial markets results from mass psychology and 'herd behaviour'. When 'news' circulates, asset prices can suddenly rise or fall, without any changes in the underlying economic conditions, such as the prices of commodities or the activities of enterprises. Financial asset prices have no real anchor. According to Keynes, the least useful economic group comprises rich owners of money capital, the parasitic 'rentiers'. They want high returns from their financial investments, otherwise they will keep their money idle and provoke the scarcity of finance for economic needs. Such behaviour has no justification, since the savings of the rich are not the result of austerity or consumption restrictions. Therefore, financial activity should be taken out of the hands of the rich owners of money capital and regulated by the state. Keynes proposed the 'euthanasia of the rentier'. The active economic agents, entrepreneurs and workers should not be dependent upon the interests of idle rentiers.

This kind of analysis is often used today by people who ask for the cancellation of the debts of less developed countries, and by those proposing a new regulation of finance (see Chapter 13). They are right to protest against the waste of labour power and real resources resulting from free financial markets. They are also right to protest against the growing inequality of income and wealth everywhere. The private regulation of exchange rates, public services and social welfare operates in favour of financial fortunes and is a source of windfall profits, instability and crisis. A new public regulation of markets and financial institutions is necessary.

However, this new regulation should involve major changes in the whole body of capitalist accumulation of wealth. Therefore, it is necessary to understand more fully the complex relationship between financial and industrial capital, and the respective roles of financiers and entrepreneurs in capital accumulation. Then we will come back to the general notion of capital, whether 'real' or financial.

According to the classical economists, all commodities, including capital goods, are created by labour in the industrial sphere. Only here is there value and profit creation. So what about finance, which is not created by labour? Marx showed how money and money capital are derived from the creation of value and the circulation of commodities (see Chapters 1 and 3). Money capital is necessary for paying wages and buying industrial equipment. It is also involved in the circulation of productive capital. Marx went further when he analysed the accumulation of capital. The exploitation of labour is the basis of capitalist profit. However, a capitalist 'credit system' is required for financing new industrial investments, and it also centralises the money of all social classes. Owners of small savings are passively involved in this process, while the ownership of financial assets is highly concentrated in the hands of a few wealthy people, including some industrial managers.

This economic role of finance does not mean that financial capital is a mere adjunct of capitalist accumulation. Even if financial returns depend on profits made in production, they have their own dynamic. Marx, after some classical economists, wrote that the evaluation of financial assets is peculiar, because they are not directly produced by labour. When we learn that, in 2001, ten trillion dollars were lost because the international stock markets had plunged, we know that this does not imply that factories and workers disappeared as the overvaluation of financial shares and wealth started to decrease. There is a depreciation only of 'paper wealth'.

But this 'paper wealth' is also a form of capital property. Financial losses do not mean that rich owners are deprived of their property rights. It is striking to see that, whatever the fluctuation of share prices, large fortunes remain concentrated in a few hands. This brings us back to the previous question: is there control of capital accumulation by the owners of large financial assets? Since the 1980s, the huge growth of joint-stock companies and investment in corporate shares has been accompanied by strong pressure by financial markets and institutions for high corporate profits and high financial rates of return from company activities. These financial standards are now the most important measure of the interests of capital. The mobility and liquidity of financial investment makes it easy to choose industrial sectors where the highest returns can be expected.

The domination of financial standards must be validated by the process of capital accumulation. Marxian analysis shows that the pressure from the owners of finance on the management of production must be relayed by industrial directors concerned with high profits from production. These directors not only have high salaries, but also obtain important share portfolios by means of stock options or in other ways. And they agree to change the organisation of industrial production in order to maximise both profits and financial rewards. *This is the objective basis for a coalition of financiers and top industrial directors.* New production and organisational standards are imposed upon small enterprises, making them more dependent on the markets for commodities and services. Outsourcing part of their productive operations, subcontracting services linked to productive activity, relocating plants to another part of the country or to other countries, all these measures are taken according to the common interests of the capitalist coalition. Even if financial and industrial capitals do have distinct features, both are involved in capital accumulation.

However, this process seems to generate a capitalism without capitalists, or a market economy driven by competition and centralisation of capitals. It looks like a natural mechanism in which the behaviour of all economic individuals is predetermined. Class divisions become invisible. So it is necessary to analyse how the class coalition of financiers and industrial capitalists opposes workers' interests.

THE CLASS COALITION OF CAPITALISTS AGAINST WORKERS

Class positions are rooted in the process of capital accumulation. According to Marx, capital is constituted not only by 'real' means of production and money, or by financial assets belonging to private owners. Since value is created by labour, capital includes a social relation of production between capitalists and workers. This determines the basic distribution of global income between profits and wages. Capital accumulation constantly reproduces the social positions of capitalists and workers (see Chapter 4).

The main features of this relation have changed in the history of capitalism. The balance of power between capital and labour is not static. Since the 1980s, it has shifted in favour of capital. Inequalities of income and wealth have increased considerably everywhere. But herein lies a paradox. The working class has suffered from the new practices of the capitalist coalition, but these were not opposed by important national or international workers' movements in the democratic capitalist countries. The working class as such seemed to disappear, broken by competition and individual interests.

Intense competition between the workers was used to cut 'corporatist' wages and employment demands. Part-time and temporary contracts were developed. Job insecurity increased for all, even when there was near 'full employment' in the United States and the United Kingdom. The mobility of workers increased. These changes exerted labour market pressure upon jobs and wages, at the expense of collective bargaining and unions. They entailed the restoration of high profits for capital.

Politics and ideology must be introduced into the analysis of class relations. Economic position and conflicts of interest are the economic basis of class divisions, but they are not sufficient to explain class struggles and compromises. Since the 1980s, neoliberal policies were not only orientated by laissez-faire, but also actively promoted new practices and social relations (see Chapters 16 and 17).

Since the 1980s, monetary policy has become the focus of economic policy. Price inflation, or the relative depreciation of money, was no longer tolerated. Whatever the origin of price increases, wages had to be stabilised and some public expenditures had to be cut. This monetary discipline was a decisive contribution of neoliberal policies to the huge development of the financial markets. While wages were stabilised, financial property was encouraged by disinflation policies.

More specific measures were at work (*Economist* 2001, pp. 3–38). In Europe and elsewhere, the privatisation of public or state enterprises promoted equities markets and property in private assets. There were also new legal dispositions and corporate strategies to finance pensions systems. There was 'a shift from state and pay-as-you-go systems towards greater emphasis on privately funded pensions' (p. 4). Mutual funds boomed. But this kind of property in financial assets by workers is indirect. Small savings are centralised by financial institutions, which drive them into markets. And even if some workers have share accounts in the enterprises where they work, they do not obtain new rights. They do not become 'associates' of the capitalists. New kinds of rewards have been developed within companies. The variable part of wages has increased. Share options, which are shares having specific rules of attribution, have become a very important part of top managers' salaries. Some US union leaders have asked for a distribution of share options to workers. They have never got it. Statistics show that a very low percentage of people in active employment get share options. And general statistics of wealth distribution show the very high concentration of financial property in few hands. There is no 'worker capitalism'.

Capitalist profitability needs not only the exploitation of labour, but also workers' loyalty to their contracts (see Chapter 5). Market discipline and new methods of production and control are insufficient. *The culture of individual opportunity* has been promoted by neoliberal policies and by the ideology of the 'new economy'. Popular access to credit for consumption goods and housing has been developed, which sustains global demand. But, since the 1980s, the access of workers to property in shares, however limited and passive it may be, was also encouraged and it has contributed to the new culture of opportunity. Collective wages were stagnant, but some workers could obtain individual compensation.

During the 1990s, the new culture was supported by US economic growth, technical progress, and booming stock markets. Capitalism's triumph looked definitive. There were crises only in some under-developed or emerging countries. Winners could afford to have compassion toward the losers, without changing their respective positions. Poverty and inequality was dissociated from class structure.

Have these neoliberal policies and the ideology of individual opportunity changed since the US crisis that started in 2000? Stock markets were hurt, profits decreased, and there have been massive layoffs in the US manufacturing sector. Unemployment has increased. The 'privileged' workers who own shares in pension funds have also been hurt by the depreciation of their meagre financial assets. The working poor have become poorer, even when they still have jobs. Their working-class identity has become more apparent, but it is still too early to grasp the consequences.

Some economists think that Keynesian policies are now coming back. The monetary policy of the US central bank has been very loose since the beginning of 2001: its short-term interest rates have been reduced from 6.5 per cent to 1.75 per cent. But this is obviously insufficient to sustain consumer demand and enterprise investments. This has led to a new fiscal policy of income tax cuts, and other measures to support demand. Does this means that 'Keynes is coming back'?

Not necessarily. As was said above, a Keynesian policy involves *a social compromise* between capitalist entrepreneurs and workers. Financial markets and 'rentiers' who own money capital should be disciplined by public rules. This was not the case for the US 'Keynesian measures' implemented in 2001. *Pragmatic neoliberals use these policies to confront emergencies.* Ronald Reagan, in 1980, adopted income tax cuts and public deficits, while breaking the workers' movement and the unions. George W. Bush, in 2001, while talking about 'compassion' with the unemployed workers, has the same strategic orientation, which does not affect the power of large companies and financial capitalists.

THE DEVELOPMENT OF A NEW ANTI-CAPITALIST MOVEMENT

The management of world business by international institutions like the IMF, the World Bank and the WTO has lost its legitimacy. In 1997–98, the IMF was unable to manage the Asian crisis. The interventions of the World Bank in the poor Southern countries have not reduced poverty. Periodic meetings of these international institutions and of the leaders of the most powerful countries (the G7) are challenged by demonstrations. The leaders of the world capitalist empire seem to be unable to provide the necessary reforms.

Since 1998, international movements against neoliberal capitalist globalisation have grown. They have exposed the iniquity of free capital movements and free trade, the privatisation of public services, the huge debts of underdeveloped countries, and the waste of natural resources. The big profits of multinational corporations are made at the expense of human beings. These critics have gained political influence since obtaining support from public opinion in the developed capitalist countries.

The new movements have developed outside the existing workers' unions and left-wing political parties. Rather than having a common political programme, they all seek human rights, including the economic and social rights crushed by the neoliberal search for profit maximisation and satisfaction of shareholders. Some of the protesters ask for business and trading ethical norms to be imposed, and for loosening the oppression of poor people and low-wage workers. Others require institutional reforms, such as control of financial flows, and real changes in the 'standards and practices' of the international institutions.

Are the critics of neoliberal capitalism and their demands for reforms a basis for a new anti-capitalist movement? Protests against the new curse of capitalism since the 1980s are an important contribution to the struggles against oppression and exploitation. Some of the proposed reforms could lead to a shift in the balance of power from capital to labour. In Italy and Germany, some workers' unions understand this and have joined the new movements. However, the rebirth of the working class is essential not only in a defensive corporatist way but also through political claims. Without the specific participation of the working class, no anti-capitalist movement can gain a victory over the coalition of financiers and industrial capitalists supported by politics and ideology.

REFERENCES AND FURTHER READING

Arrow, K.J. (1981) 'Real and Nominal Magnitudes in Economics', in D. Bell and I. Kristol (eds.) *The Crisis in Economic Theory*. New York: Basic Books.

Bellofiore, R. (1998) (ed.) *Marxian Economics: A Reappraisal (Essays on Volume 3 of Capital)*. London: Macmillan.

Economist (2001) 'A Survey of Global Equity Markets', May 5.

Keynes, J.M. (1936) *The General Theory of Employment, Interest and Money*. London: Harcourt Brace Jovanovich, 1964.

Marx, K. (1867) *Capital*, vol. 1. New York: International Publishers, 1967.

Sweezy, P.M. (1994) 'The Triumph of Financial Capital', *Monthly Review* 46(2) pp. 1–11.

11 War, Peace and Capitalism: Is Capitalism the Harbinger of Peace or the Greatest Threat to World Peace?

Christopher Cramer

One of the political arguments for capitalism has always been that it could tie people up with the relatively benign business of money-making, thus diverting them from the more nefarious activities of seeking power and making war, to which they might otherwise be prone (Hirschman 1977). It is still often presumed that capitalism is pacific, because it knits people together within and among countries in the bustle of production and exchange, consuming their attention and raising the costs of war. A very different idea of the properties of capitalism is captured by Wood: 'I am convinced ... that capitalism cannot deliver world peace. It seems to me axiomatic that the expansionary, competitive and exploitative logic of capitalist accumulation in the context of the nation-state system must, in the longer or shorter term, be destabilising, and that capitalism – and at the moment its most aggressive and adventurist organising force, the government of the United States – is and will for the foreseeable future remain the greatest threat to world peace' (1995, p. 265 – see Chapter 9). This chapter discusses whether there is a clear Marxist position on war or on the links between war and capitalism. It then shows the consequences of not adopting a historical political economy perspective. It argues for the relevance of a historically minded analysis of contemporary war in which the role of capitalism – advanced and nascent – is central but complex.

IS THERE SUCH A THING AS A MARXIST THEORY OF WAR?

Marxist theory is concerned with social conflict, with crisis, and with the commonplace brutality of social relations in many circumstances. The Marxist analysis of capitalism highlights the founding violence of this form of social organisation through primitive accumulation. After the establishment of capitalist relations of production there remain inherent tendencies towards occasional crisis. And a Marxist analysis also stresses the significance of class conflict, where exploitative relations are bound to generate antagonism of various forms. Further, classical Marxist ideas of how capitalism could be supplanted by a different and even more progressive mode of production sometimes revel in imagery of violent conflict. For all this, however, there is not much – certainly in Marx's own writings – that directly explains war or the relationship between war and capitalism. This is so despite the fact that Marxist theory was elaborated at a time characterised by major civil wars (including the American Civil War) and international wars, and despite the fact that both Marx and Engels were attentive readers of Clausewitz.

However, there are components of original Marxist thinking that suggest some perspectives on war and capitalism that might be absorbed into a historical political economy of contemporary conflicts. To begin with, Marx saw war as archaic. Marx, recall, waxed rhapsodic about the historically transformative powers of capitalism. Capitalism was revolutionary in its progressive consequences for human society. One of the senses in which capitalism was superior to any previous set of social relations was precisely that it was not a system of perpetual warfare in the literal sense. For medieval European society more or less had been that – a society in which war was the dominant institution and peace merely an occasional interlude. This was an Enlightenment insight: for arguably peace was only invented during the Enlightenment as a serious prospect, and there were considerable hopes that, having overthrown the institutions of the *ancien régime*, societies would be able to live without war.

If war as a central institution of society was made archaic by capitalism and bourgeois society, nonetheless war could play a role in the success of capitalism. In several places Marx writes that social relations as they take shape in war and military organisations can accelerate the development of the productive forces. 'In general, the army is important for economic development. For instance, it was in the army that the ancients first fully developed a wage system ... This division of labour within one branch was also first carried out in the armies. The whole history of the forms of bourgeois society is very strikingly epitomised here' (quoted in McLellan 1977, p. 342). Providing for armed forces has often generated innovations that have then spread through societies. Famously, the concept of sizes and ready-to-wear clothing developed during the American Civil

War from the need for uniforms. Indeed, capitalism, war and modern nation states fed off one another in an extraordinarily expansionary combination from the seventeenth century onwards (Tilly 1990). Capitalism has a distinctive technological dynamism. Military demand, the compulsion of war, and ideological urgency in the perceived threat of war have harnessed this dynamism with dramatic effect. Since the early days of capitalism arms manufacturers (in the seventeenth century especially British, Dutch and French) competed for export markets in Europe, America, and elsewhere, including Africa where the arms trade was integral to the slave trade and where this trade revolutionised warfare. More recently, of course, the appliance of capitalism's technological concentration to military ends has fuelled a phenomenal arms race and, especially since the end of the Cold War, a proliferation of industrial production of arms internationally.[1] A nice example of the power of the interests of arms-oriented capital overriding liberal idealism was the announcement, in December 2001, that the British government was planning to approve an export licence for a military air traffic control system (costing well over the average for civilian systems) to Tanzania, one of the poorest countries in the world.[2] This is but one example of the persistent links between states and military productive interests, a set of linkages that used to be taken as so powerful that it dominated capitalist economies and became known as the 'military industrial complex', or MIC. The MIC idea has rather faded from view, as indeed has the argument that military expenditure and war-related production might be positively necessary to the survival of the capitalist economy.

One way in which some Marxists have viewed capitalism as especially amenable, at least, to war is through the development of underconsumption theory. From this perspective, common particularly in the 1970s, capitalism is prone to crisis when its reliance on the exploitation of labour contradicts the need for sufficient demand for commodities produced in capitalist relations of production. Military production – and the expansion of an MIC, driven by state procurement – has sometimes been seen as an inbuilt mechanism within capitalism of defence against underconsumptionist crises. To the extent that military preparedness requires realisation in war to justify continued investment, then this argument would support a pro-war tendency within capitalism. At the very least, one could argue that the strength of capital tied up in military production and provisioning contributes greatly to the shaping of foreign policy and the way wars are fought. For example, US military technology has evolved a particularly strong commitment to air power; and US military commitments in, for example, Afghanistan in 2001, have seemed to some critics to rely inordinately on air power. However, there has been strong criticism of the underconsumptionist position, chiefly on the grounds that in jumping straight from a very abstract theory to instant explanation of empirical facts it provides no mediating links, that its underlying theory of capitalist crisis is over-wrought, and that it is completely arbitrary to assign exclusively to the military the potential for moderating tendencies to crisis in capitalist economies (see Chapters 9 and 15).[3]

Nonetheless, there is a final sense in which capitalism might well inherently support the likelihood of violent conflict. For capitalism is by its very nature conflictual: the logic of desperate competition that compels capitalists – especially perhaps when framed within nation states and the organisation of national interest – could be expected to generate regional and international violence, including violence in the form of war. This tendency might be mitigated, however, by the fact that capitalism is transnational and develops complex interconnectedness between people. The key, however, is to see the independence of the nation state as the principal unit of political organisation and international legitimacy, shaping capitalist competition into a potentially lethal form. There might not be anything inherently warlike about capitalism, but when it is harnessed to national power and competition it easily becomes so. Here the role of the French government, in particular, in backing and arming the Habyarimana regime in Rwanda that unleashed genocidal violence in 1994 against (mainly) Rwandan Tutsis is an obvious example. So too is the rather more complex US involvement in the Middle East as part and parcel of the Israeli–Palestinian conflict and also the wider social and political conflicts that bred al-Qa'ida terrorism from the late 1990s onwards. Here it is not simply 'the national state' but the configuration of political influence on the US state that has held together much of the Middle East in a vice of artificial stability through support to regimes in Egypt and Israel, among others.

For surely it is the combined causal powers of capitalism and national interest (along with a range of other material and ideological factors) that are realised in arms races and military engagements among capitalist nations. Similarly, it is the combined causal powers of the transition to capitalism (the prolonged, traumatic 'moment' of change), the non-linear history of state formation, the diverse roles of non-national collective identities, and the interests of international capital that are realised in the 'civil wars' of the world in recent years. Thus, while there has often been a tendency to insist on the total subordination of war and militarism to capitalist production logic, this is never analytically very successful. Marx and Engels themselves ultimately portrayed war as 'a relatively independent variable in the ever-changing human scene' (Gallie 1978, p. 79).

Further, although the causal powers of capitalism are central to any understanding of modern conflict, these cannot be 'read off' effortlessly from some rigid logical schema. Rather, all the ambiguities of capitalism are revealed in the relationship between capitalism and violent conflict. Capitalism can mitigate the proclivity to war just as it can in different circumstances provoke war. Capital tends to require peace but often thrives on conflict. The expanding domain of capital may knit together different peoples in a common association around the possibilities and political challenges of bourgeois society; but at the same time the international reach of capital may raise the stakes of local conflicts. Obvious examples of this last tendency include the markets for high-value primary commodities that play a central role in sustaining and scaling up conflicts around the world: markets for diamonds; the vast set of linked economic activities relying on oil; or the mobile phone, games console and space technology industries and their hunger for coltan, the heat-resistant metal whose deposits are concentrated in the Kivu districts of eastern Congo (Kinshasa) and whose extraction is fought over by contending Congolese, Rwandan, Ugandan and other regional groups.

There is an accommodation between capitalism and other factors, such as the nation state, and a social propensity to violence, that is independent from specific historical epochs. Capitalism itself remains of central significance, however, as of course does the scope within capitalist society for political struggle to have real consequences. From this perspective, we should be wary of analyses that repeat in reverse the errors of some strains of radical analysis. For example, Shaw (2000) suggests that the institutions of war fed on capitalism and, once sufficiently engorged, squashed and subordinated the capitalist mode of production. This to some extent is what E.P. Thompson argued when he claimed there was in modern industrial society a 'logic of exterminism'. For all those enamoured by the attractions of Western liberal values and the defence of 'civilisation', it is certainly salutary to recall the thread of extreme and mass viciousness in modern Western society. Modern exterminisms were prefigured in Montaigne and Swift (for example, in *Gulliver's Travels*, in the Houyhnhnms' debate on whether to exterminate the entire population of crude, human-like Yahoos). European genocidal tendencies, far from being the exclusive preserve of Nazi Germany, arguably have been more general and are rooted in imperial ideology and crude social Darwinism. These tendencies were realised not just in the Holocaust but before then in colonialism and were captured perfectly in a single phrase spoken in Conrad's *Heart of Darkness* by Kurtz: 'Exterminate all the brutes' (Lindquist 1997). However, the human imagination of violence has been a rich one for centuries and for centuries it has been realised horrifically. There is little real evidence of a peculiarly modern social death wish. Arguably, what we see in genocide, in the nuclear arms race, in ethnic cleansing and so on is less a logic of exterminism and more some of the particularly morbid forms in which dynamic capitalism in advanced countries and insecure capitalism in developing countries manifests its linkages with competitive collective identities and a historically entrenched human violence.

A related discussion is that among sociologists concerning whether 'war and violence are parts of modernity and not only of its genesis' (Joas, quoted in Roxborough 1999, p. 494). Roxborough's answer is that war could well persist as part of modernity for three reasons: conflicting values (since peaceable consumerism and commercialism might not be the sum total of value sets possible within modernity); aggregation problems (the ways in which individually rational decisions and preferences become aggregated through institutions might then favour a clash between particular interest groups outweighing majority wants, or, simply, there may be miscalculation leading to blundering into war); and limitations on rationality (cognitive frameworks or ideologies may lead people to misunderstand their predicament and go to war as a result). Shaw goes further: rather than stating that war remains an ongoing possibility given the shortcomings of modern rationalism he argues that the genocidal tendency in recent wars (e.g. in Rwanda and the former Yugoslavia) shows precisely that 'in modernity, *war is the problem*' (emphasis in original).[4] The focus of these contributions seems to me misplaced. In Shaw all historical ambiguity is lost under the weight of the contemporary 'mode of warfare' and the logic of exterminism. Capitalism is entirely subsidiary, simply a contingent and enabling factor. Meanwhile, Roxborough sustains a sense of ambiguity in modernity but excludes the material and the dynamics and tensions of capitalism as central to that historical ambiguity vis-à-vis war. Neither approach adequately captures the relational content of social, economic and political conflict that is implied more effectively in Marxist traditions of political economy.

In short, if we combine the interest in the origins of capitalism and the underlying relations sustaining it with a recognition of the ongoing propensity to contradiction, crisis and competition, with an awareness of the historical independence of violent conflict from specific material epochs, and an awareness of how those specifics nonetheless shape and carry the baton of war, then we have a powerful basis for understanding contemporary conflicts.

WHAT ARE THE CONSEQUENCES OF NOT TAKING A MARXIST PERSPECTIVE ON WAR AND VIOLENT CONFLICT?

The flip side of the 'logic of exterminism' school of thought is liberal optimism. Most contemporary analyses of war fall in the liberal camp. Contrary to Marx's 'tragic view of history', the liberal interpretation of war and peace is based on the assumption that all good things go together. Economic progress and political progress are mutually supportive and enjoy an entirely uncomplicated relationship. 'The liberal dream which stemmed from the Enlightenment project was that the modernisation of society would lead to the disappearance of war', as Roxborough puts it (1999, p. 491). Naturally, therefore, the persistence of war must mean that there has simply not been enough modernisation.

Again reflecting a lack of capacity for ambiguity or complexity, the 'liberal interpretation of war' inherited from the nineteenth century considers war to be always and exclusively negative in its consequences. Although there were important empirical challenges to this stance early in the twentieth century, the frame of mind survived and reappeared in various forms, including exercises working out the total economic costs of war in developing countries. If there is a liberal interpretation of war, there is also a liberal interpretation of peace and the transition to peace. According to this position, Western democracy is self-evidently a 'good thing', as are NGOs, adjustment to a more market-based economy, and so on; and the relationships among these factors is also unproblematic. Hence the model for peace is to nurture governments that commit to structural adjustment policies and strongly to encourage a swift introduction of procedural democracy in the form of 'free and fair' elections.[5]

So, war is a terrible thing that arises from lack of modernity and makes things worse, always. This often translates into the causal connection presumed by many between poverty and war. Poverty causes war and war causes poverty. Another dimension of the liberal perspective on violent conflict is the mentality of collapse. For most such analyses argue or presume that wars in developing countries are a function of collapse and reversal – of the state, of modernisation, of development. Wars, particularly the post-Cold War conflicts, are commonly seen from this view as apolitical, untouched by ideology but rather driven by base greed and/or a social retreat into conflicts of ethnic animosity inherited through some process of (assumed) social Darwinism. There is little scope from this perspective for inquiring whether conflicts might be *part of* a tumultuous and long process of state formation and the establishment of capitalism, not just a threat to that process; little scope for seeing that although all war is sickening some wars might nonetheless have progressive consequences.

A particular and more formal variation of the liberal perspective on war is built on the axioms and institutional influence of neo-classical economics. According to this approach, civil war is the outcome of rational choices of individuals seeking to maximise their utility and faced with a trade-off between co-operation and conflict. Conflict will be chosen under certain circumstances that determine whether or not conflict is more profitable at the margin than co-operation. The most common factor tipping the scales of choice towards conflict is poverty. For, it is claimed, the poor have a 'comparative advantage in violence': this is because they have next to no other opportunities, therefore the opportunity cost of engaging in violence is close to nil. Models along these lines are, however, purely abstract and speculative until there is some effort to make them empirically operational. Some effort has been made to do this, for example, by claiming that 'greed' rather than 'grievance' explains the incidence of civil wars.

The trouble is that empirical applications of these models are unsuccessful. The empirical tests do not perform very well. They are constructed from data whose reliability and comparability are highly questionable. And they are poorly designed: both because the samples are sometimes biased and because the variables used as proxies (because they can in principle be quantified) for more direct concepts identified in the abstract models do not correspond neatly to their theoretical counterparts. A high share of primary commodity exports in total GDP, a preponderance of young males in the population structure, and low average years of schooling are taken in one model (Collier 2000) to signal the influence of 'greed' before being tested for correlation with the incidence of civil war. Yet it is equally plausible that strong informational content for these three variables may indicate, instead, widespread social frustration and 'grievance'. Alternatively, such a conjuncture (lots of young, uneducated and therefore [sic] unemployed males surrounded by primary commodities) might be the starting point to explore the complex interaction between, or joint determination of, greed and grievance.

Varieties of liberal analysis are all unsatisfactory and restrictive. They treat the material dimensions of war as fetishes: giving magical causal powers to degrees of resource concentration or fixing the determinants of conflict in optimal combinations of poverty, demography and other variables. Efforts to incorporate the social have no historical or relational content. Furthermore, any pretence to capture human agency through the incantation of rational choice is belied by a staggering determinism: the choice is always made, written in the econometric stars. However elusive the understanding of war might ultimately be, it has to address the material, it must be historical and relational (what is conflict if not relational?) and it has to allow for human actions and policy decisions. Surely, also, a useful analysis of conflict must contain some focus on historical change or transition. Only an analysis rooted in a Marxist tradition can hope to meet these demands.

A BRIEF ILLUSTRATION: WAR IN ANGOLA

The briefest illustration of the war in Angola – a country at war more or less constantly over the past 40 years – helps to tie together the themes of this chapter. Despite first impressions, Angola does not neatly fit the template analyses currently on offer. Resources play a role in the war, but the conflict is not simply produced by oil and diamond abundance, though it is indeed reproduced through this abundance. The instrumental use of violence by greedy elites is a characteristic of the war but can only be appreciated against a more subtle history of power struggles and grievances. Various forms of sub-national collective identity have helped shape the divisions in the war; however, to claim to 'explain' this war by reading from a score of ethnic fragmentation would be laughable. Angola's is a 'civil war' that has been fuelled by external interests throughout. And so it is an international war (now as it was during the Cold War), but it has never been a 'proxy war' pure and simple. Furthermore, Angola's current spate of warfare bears some of the features of what are often called 'new wars'; yet it cannot be understood as anything other than a war with old and enduring roots.

It is only possible to begin to make sense of war in Angola through a historical analysis of capital and of clashing class interests and wrenching experiences of class formation, from the moment the Portuguese arrived in the late sixteenth century to the present day. Such an account would bind together: the consequences of initial encounters with merchant capital fuelling the slave trade and with industrial capital generating technical innovations in and exports of guns from Europe; the organisation of the colonial economy, including the ways in which differences in the spread of capitalism overlapped with the beginnings of distinct zones of Angolan nationalism (shaped by domains of different mission groups and their schools); the way that the ferocity of the Cold War scaled up rival conflict among anti-colonial groups; the logic of oil and diamond markets; and the way in which foreign companies currently seem to be acting as vehicles of, and influences on, the foreign policies of major powers.[6] Obviously, wars like Angola's owe a great deal to contingency, to national interest, to local specificity; but equally obviously they are driven by the compulsive logic of capital.

Social transformation and state formation in Angola have been extraordinarily disruptive and drawn out and remain incomplete. Given its own history, and against the background of European history, this is hardly surprising. It might also be noted that 'peace' in Angola – where vast numbers of people have been forced away from rural subsistence livelihoods by the privations of war – will not bring an end to this process and its brutality. For all the well-meaning policy advice that will be meted out during a peace process and its aftermath, about 'reconstruction' and so on, it may be expected that the accumulation of land and other assets, primitive accumulation to be sure, will dominate the real politics of Angola and will – if other 'post-conflict' experiences (in Nicaragua, Mozambique, El Salvador) are any guide – continue to be characterised by violence.

CONCLUSION

The analysis of war shows capitalism at its most ambiguous. It might even be that there is this paradox: that capitalism is actually in some ways more pacific than most other known forms of social organisation, but at the same time many of its qualities lend themselves better than other modes of production to increasing the intensity of conflict. This chapter has shown how war or social violence and the transition to capitalism are commonly bound together. It has also shown how in the contemporary world economy there is a distinctive binding of this traumatic transitional 'moment' and the interests of advanced capitalist nations, consumers and enterprises. This kind of analysis should make one wary of the pretty prognostications of liberal theory and policy advice. And it should alert one to the possibilities for horrendous conflicts to be associated with progressive outcomes: if this is the case there is a need to look for where those outcomes might emerge from in war and how to promote their manifestation. The analysis has also shown that war is likely to remain a feature of our world.

REFERENCES AND FURTHER READING

Collier, P. (2000) 'Doing Well out of War: An Economic Perspective', in M. Berdal and D. Malone (eds.) *Greed and Grievance: Economic Agendas in Civil Wars*. Boulder and London: IDRC/Lynne Rienner.

Gallie, W.B. (1978) *Philosophers of Peace and War: Kant, Clausewitz, Marx, Engels and Tolstoy*. Cambridge: Cambridge University Press.

Hirschman, A.O. (1977) *The Passions and the Interests: Political Arguments for Capitalism before its Triumph*. Princeton: Princeton University Press.

Kaldor, M. (1999) *New and Old Wars: Organised Violence in a Global Era*. Cambridge: Polity Press.

Lindquist, S. (1997) *'Exterminate All the Brutes'*. London: Granta Publications.

McLellan, D. (1977) *Karl Marx: Selected Writings*. Oxford: Oxford University Press.

Pythian, M. (2000) *The Politics of British Arms Sales Since 1964*. Manchester: Manchester University Press.

Roxborough, I. (1999) 'The Persistence of War as a Sociological Problem', *International Sociology* 14(4), pp. 491–500.

Shaw, M. (2000) 'The Contemporary Mode of Warfare? Mary Kaldor's Theory of New Wars', *Review of International Political Economy* 7(1), pp. 171–80.

Smith, R. (1977) 'Military Expenditure and Capitalism', *Cambridge Journal of Economics* 1(1), pp. 61–76.

Tilly, C. (1990) *Coercion, Capital and European States, AD 990–1992*. Oxford: Blackwell.

Wood, E.M. (1995) 'Capitalism and Human Emancipation: Race, Gender and Democracy', in *Democracy Against Capitalism*. Cambridge: Cambridge University Press.

NOTES

1. Engels captured the anxious peace of the arms race well: 'Peace continues only because the technique of armaments is constantly developing, consequently no one is ever prepared; all parties tremble at the thought of

world war – which is in fact the only possibility – with its absolutely incalculable results' (Gallie 1978, p. 92).

2. On the politics of the British arms trade, see Pythian (2000).

3. On the economic consequences of military expenditure see Smith (1977).

4. Shaw's argument is developed around the concept of 'new wars', coined by Kaldor (1999).

5. Even this involves turning a blind eye towards manipulation by international managers or at the very least towards some of the shortcomings of election conduct in Cambodia, Bosnia-Herzegovina, Mozambique, etc.

6. Through the Cold War period, US oil companies tempered American hostility to the MPLA government with whom they did a roaring trade; more recently, French government officials have been allegedly embroiled in corruption scandals including Elf Aquitaine's interests in Angola.

12   Understanding Capitalism in the Third World

Elizabeth Dore

How Marxists understood capitalist development – or its absence – in Latin America, Africa and Asia changed dramatically over the course of the twentieth century. Broadly speaking, in the first half of the century Marxists believed that the countries of Latin America, Africa and Asia were underdeveloped because they *were not* capitalist. In the second half of the twentieth century most socialist scholars turned this interpretation on its head and argued that tricontinental countries were underdeveloped because they *were* capitalist. This radical revision was part the result of changing socio-economic conditions, and part the product of changing revolutionary strategies in the Third World. This chapter examines ruptures as well as continuities in Marxist approaches to underdevelopment and shows that how we understand capitalist development matters.

FEUDALISM AND THE OLD LEFT

Before the 1960s, the Old Left espoused the view that underdevelopment was the result of the absence of capitalism. From the 1920s to the 1940s, the official position of the Comintern (also called the Third International – the association of Communist Parties founded by the Bolsheviks to promote worldwide socialist revolution) was that all colonial and post-colonial countries were feudal. Feudal, in this sense, meant societies ruled by landlord and merchant classes whose power and wealth derived from exploiting the peasantry by overtly violent, non-market means (see Chapters 1 and 9). In this line of argument, feudal ruling classes in alliance with imperialists – the capitalist classes in industrialised countries – blocked the kind of economic competition that gives rise to technological advances and modern market societies. Therefore, according to the Old Left, because the colonies and post-colonies were not-yet-capitalist, or pre-capitalist, they lacked the preconditions for socialism. Consequently, the Comintern's mandate for communists and their allies in underdeveloped countries was to pursue a 'two-stage' revolutionary strategy. To hasten their country's transition to capitalism, they should first promote anti-feudal and anti-imperialist struggles. Then, after the capitalist transition, 'the national bourgeois revolution', socialists should redirect their efforts to anti-capitalist revolution. In this scenario, before their countries were capitalist, peasants, workers and other subordinate classes would form tactical alliances with those sectors of the local capitalist class – the national bourgeoisie – that had conflicts with, ergo were oppressed by, the imperialists and their feudal allies. In stage two, after the feudal classes were overthrown and the imperialists weakened, revolutionary peasants and workers would shed their erstwhile bourgeois allies and struggle for socialism. While this summary of the Comintern's 'line' on Third World revolution is schematic, to my mind the policy prescriptions tended to be schematic too. However, most major communist leaders, including Lenin, Stalin and Mao Zedong, advocated some version of this explicitly linear scenario of two-stage revolution.

Adherence to the Comintern's prescriptions led communist parties and radical activists in Latin America, Africa and Asia into some strange pacts. Especially at the time of the Second World War, when pursuing both the two-stage revolutionary strategy and the Comintern's United Front Against Fascism, communists ended up supporting an assortment of unsavoury dictators and repressive regimes. For instance, communists in Nicaragua backed Anastasio Somoza, and the Cuban Communist Party supported Fulgencio Batista long after he turned from reformism to reaction.

THIRD WORLD CAPITALISM AND THE NEW LEFT

In the 1960s, radicals turned this scenario for Third World revolution upside down. In Latin America, the revolution from below began when Fidel Castro and Ché Guevara led an armed uprising against the dictatorship of Batista, Washington's lackey in Cuba. Before and after the triumph of the Cuban Revolution in 1959, Castro and Guevara cast aside reformist politics associated with Latin America's communist old guard: they advocated socialist revolution throughout the Third World. Although the Cuban leaders' political analysis was rooted in the country's special history of rapid capitalist development, led by US-owned sugar companies, Castro and Guevara applied this thinking to all countries of the Third World. To this end, the Cuban government forged a new International, the 'Tricontintental Congress' bringing together revolutionaries from the post-colonial world. Notwithstanding Cuba's reliance on Soviet economic and political support, the Castro government played a leading role in challenging the politics of the Old Left, which remained the official 'Moscow line' for revolution in the Third World.

A combination of events set in motion political upheavals in the 1960s: anti-colonial struggles in Africa, Asia and the Caribbean, the civil rights movement in the United States, and Washington's war in Vietnam. In this era, the influence of the Cuban Revolution should not be underestimated. Castro and Guevara inspired revolutionaries around the world to found parties that regarded the Cuban Revolution as their model. In 1968, student revolts sparked rebellions in other sectors of society against the Vietnam War and against white supremacy in the United States, against capitalism in Western Europe, and against Soviet-style socialism in Central and Eastern Europe. Large demonstrations in the United States, Paris, London, Mexico City, Rio de Janeiro, Rome, Berlin, Prague, Warsaw and Tokyo – some verging on uprisings – manifested widespread rejection of the status quo and disaffection with the politics of the Old Left.

These upheavals gave birth to a New Left. Following in the footsteps of Castro and Guevara, the New Left repudiated Old Left politics, especially its vision of the Third World. Radicals in developed and developing countries virtually swept away the doctrine that neocolonial countries were feudal and lacking the pre-conditions for socialist revolution. They turned this thesis upside down and argued that the countries of Latin America, Africa and Asia had long been capitalist, and that capitalism – not its absence – caused underdevelopment. In this revision of Third World history, capitalism developed in distorted forms in post-colonial countries and conditions there were particularly ripe for socialist revolution.

The shift in Third World revolutionary politics was a tsunami of major proportions. In the broadly Marxist debate of the 1960s and 1970s about the causes of underdevelopment, three schools of thought stand out. Dependency theory quickly became the orthodoxy in the field (Frank 1969), followed by World Systems theory (Wallerstein 1979), a variation on the dependency theme. In this theoretical current, capitalism is – first and foremost – a system of international exchange in which 'metropolitan' or imperialist countries appropriate 'surplus' from colonial or neocolonial countries. The major conclusions of this theory are (a) that European trade took capitalism to the far reaches of the colonial world, where it took root as long ago as the sixteenth century; and (b) that capitalism – understood as surplus appropriation among countries through international trade – caused underdevelopment in colonial countries and development in imperialist countries. Dependency remained the prevailing wisdom in Third World Studies for several decades. It left behind as one of its most enduring – and to my mind troublesome – legacies the almost unquestioned view that colonial and post-colonial countries have been capitalist for a very long time – at least since contact with Europeans.

A different approach to understanding capitalism in the Third World was grounded in more traditional Marxist methods. Unlike dependency theorists with their almost exclusive focus on exchange between countries, Marxist scholars viewed class relations – a relationship rooted in processes of production – to be the motor force of capitalist transitions. To understand development and underdevelopment, Marxists analysed the social forces that promoted and prevented transformations to free wage labour, particularly in agriculture (Brenner 1977). This approach focused on how large landowning classes appropriated surplus products and surplus labour from the people who directly worked the land. Overall, Marxists sought to understand how and when peasants became proletarians, and how these processes affected technical change and economic growth (Weeks and Dore 1979).

The Marxist literature on development and underdevelopment veered away from dependency theory in a number of crucial ways. Marxists tended to hold a more contradictory view of the history of capitalism – its uses and abuses – in the Third World. In the Marxist framework, capitalism rests on exploitation in production: on the propertied classes' appropriation of the labour (or labour power) of people who own no property and who, therefore, have to sell their labour power to survive. Because capitalism engenders competition among capitalists over profits, inter-capitalist rivalry tends both to increase the exploitation of workers and to drive forward technical change. In short, capitalism promotes growth and development – *of a particular sort*. Consequently, competition among capitalists creates conditions for – or the possibility of – improving workers' standards of living (see Chapter 4). Whether or not this occurs depends on workers' struggles against capitalists, not on some technical fix inherent in capitalist production. In sum, whereas dependency theorists see capitalism as an unmitigated evil, most Marxists see capitalism as an evil that rests on class exploitation and political subjection, but mitigated in so far as capitalism contains within it a drive to raise labour productivity. In capitalist societies increasing labour productivity tends to be harnessed to intensifying exploitation; however, technical change, resulting in higher productivity, potentially liberates humankind from some of the drudgery of work. But emancipation from toil – even in this narrow materialist sense – is more of a possibility in post-capitalist societies than within the belly of the capitalist beast. From this perspective, Marxists regard some aspects of capitalist development in the Third World as *exploitative and progressive*: using 'progressive' not in the sense of good, fair or just, but in the sense of creating conditions for economic advancement that might promote human liberation, instead of exploitation, in a socialist order.

A related difference between dependency theory and Marxist theory is that in the former, Third World countries have been capitalist for something in the order of five hundred years, while in the latter capitalism has had a more recent history in the post-colonial world. Marxist scholars frequently emphasised the heterogeneous and zigzagged nature of capitalist transformations in Latin America, Africa and Asia (Cooper et al. 1993). Nevertheless, in my view, swayed by the current in Third World history, some Marxist writers tended to overemphasise the capitalist nature of agrarian change and to minimise the staying power of non-capitalist relations. Rather like explorers on the lookout for the earliest sightings of free wage labour, they might have been predisposed to making a discovery. As a consequence, I think there was a tendency even within the Marxist tradition to present a great variety of different kinds of social upheavals as *the* capitalist transition, or as major turning points on the capitalist road.

The last body of broadly defined Marxist writings on the rise of capitalism in the Third World was a kind of 'third way'. A school of thought known as 'articulation of modes of production' sought to meld together elements from dependency theory and traditional Marxism (Foster-Carter 1978). Writers of this persuasion emphasised the ways that capitalism coexisted with non-capitalist social forms. Overall, they argued that non-capitalist class relations persisted only (or mainly) because capitalists appropriated surplus labour and/or products from peasants. Ergo, non-capitalist class relations survived into modern times only when and if they played a functional role in capitalist development.

The broadly defined Marxist debate of the late twentieth century about the causes of development and underdevelopment was exciting. To understand Third World capitalism and anti-capitalism in our time we would do well to read – or reread – the literature at the heart of those international controversies. In these debates, scholar/activists tried to understand the world in order to change it. With the benefit of hindsight, but at the risk of flattening out a rich and fertile field, I propose that notwithstanding key differences in the three schools of thought, taken together they tended to portray the Third World as capitalist, or as far advanced on the capitalist road, a road with few detours or byways. Once the political tide had turned from the Old Left to the New, the view that colonial and neocolonial countries were not capitalist (that they were 'feudal'), or that they retained significant non-capitalist elements, fell into disrepute.

THEORY AND PRACTICE: THE SANDINISTAS' HISTORY

Leaders of anti-imperialist movements in Latin America, the Caribbean, Africa and Asia were at the forefront of the new-wave history of the Third World. In each epoch the call for change adapts itself to the radical rhetoric of the time. It is not surprising that in the 1950s and 1960s, anti-colonial leaders used the language of anti-capitalism and professed a commitment to socialism. Independence struggles were pitted against the world's capitalist powers; for this reason, the language of socialism seemed to lend itself more readily to the discourse of anti-colonialism than the language of feudalism and capitalism. Yet, with few exceptions, after independence new governments in Asia, Africa and the Caribbean fostered capitalist development.

Following the Cuban Revolution, Latin American radicals also adopted the language of socialism – even in countries seemingly ill suited to anti-capitalist or post-capitalist movements. With communism the enemy of the US government, it is unsurprising that throughout Latin America anti-imperialism was framed in the ideology of the enemy's enemy: Marxism. The Frente Sandinista de Liberación Nacional (FSLN) in Nicaragua grasped the transformative power of the new Third World history. Many leaders of the FSLN wrote history; they reinterpreted the past with the explicit objective of inspiring their compatriots to revolutionary action. From its formation in 1961, the FSLN disseminated its vision of Nicaraguan history through pamphlets, songs and speeches. After a popular armed insurrection defeated the Somoza dictatorship and brought the FSLN to power in 1979, the Frente Sandinista created a historical institute whose mission was to remake 'official history' in the image of the Sandinista vision of the past.

The FSLN's re-vision of the past portrayed Nicaragua as a country of rebellious rural proletarians that was ripe for socialism. The Frente's founder, Carlos Fonseca, initiated the new history, retelling the national story as a sequence of popular uprisings against US imperialism. Fonseca's history recovered Augusto Sandino, the leader of a guerrilla peasant army that fought the US occupation of Nicaragua from 1927 to 1933. Where Somocista history had banished Sandino the 'rural bandit', Sandinista history resurrected Sandino and portrayed him as the 'saviour of the nation'. In Fonseca's interpretation of the past, Nicaragua was a nation of revolutionary worker peasants who repeatedly, and against all odds, resisted US intervention. In this version of the past, Sandino was both father of the nation and the embodiment of the Nicaraguan national character. Nicaraguans could fulfil their national destiny by following in the footsteps of Sandino and of his direct successors, the Frente Sandinista de Liberación Nacional.

Jaime Wheelock, leader of the Proletarian Wing of the Frente Sandinista, developed the Sandinista school of history into the leading current in Central American revolutionary thought. Wheelock, an accomplished historian, said he wrote his most important book, *Imperialismo y dictadura: Crisis de una formación social*, to persuade party militants that their struggle was not simply anti-dictatorial and anti-imperialist, but anti-capitalist (1979, p. 12). In Wheelock's interpretation of the past, Nicaragua possessed all of the preconditions for socialist transformation. Its capitalist transition occurred at the end of the nineteenth century, and by 1960 national capitalism had developed to a mature stage. Wheelock argued that Nicaragua was a country of rural proletarians who, because of their class position and political consciousness, would join an anti-capitalist revolution.

Subsequently, a number of historians argued against the Sandinista view of the past, saying it reflected their political ideology and presented an inaccurate interpretation of history. In the counter-interpretation, Nicaragua was a pre-modern society of landlords and peasants as recently as the 1940s and 1950s. And, although capitalism developed rapidly in the 1960s, class consciousness changed slowly and the world-view of rural people tended to be more backward looking than forward looking. Rural workers violently evicted from land in the previous decades retained a deep longing for landholding. Consequently, rather than developing a working class or socialist consciousness, poor Nicaraguans aspired to return to what they viewed – somewhat romantically – as their traditional peasant way of life. In the counter-history, Nicaragua had a small, highly unorganised working class, a weak and fragmented communist movement, and scant tradition of socialist thought in the twentieth century. In short Nicaragua was not, as Wheelock argued, ripe for socialist revolution (Dore, 2003).

Whatever the merits and demerits of Sandinista history, unlike their rivals from right to left across the political spectrum, only the FSLN was successful in leading a powerful movement against the dictatorship. The Sandinistas' success rested on their ability to galvanise the imagination of the masses of Nicaraguans. This they achieved, in part, by rewriting history. Paradoxically, the anti-capitalist discourse that played a role in inspiring Nicaraguans to take up arms against the dictatorship, when put into practice in the countryside provoked large numbers of peasants to take up arms again: this time against the Sandinistas and their revolution.

Not surprisingly, after Wheelock became the Sandinista Minister of Agriculture, his vision of history strongly influenced state policy in the agrarian sector. Despite peasant demands for land, the FSLN refused to distribute land to the tiller. Wheelock declared that because for several generations the majority of the agrarian poor were rural proletarians, and because capitalist class relations had predominated in the countryside for over a century, distributing land to the peasantry would be a retrogressive step.

The Sandinista Agrarian Reform created large state farms and the government promised to deliver on the classic demands of rural workers: improved wages and working conditions. In line with socialist ideals, the Ministry of Agriculture set about encouraging workers' participation – if not control – over what the FSLN hoped would become huge, high-tech farming complexes. In the event, large numbers of rural Nicaraguans opposed the Sandinistas' state-centred agrarian policy and continued to press for land distribution to peasant households. When the Sandinista government ignored peasant demands, many rural people joined the Contras, the armed opposition founded and funded by the US government to overthrow the Sandinistas (Dore and Weeks 1992).

By the middle 1980s it became evident that the Sandinistas' vision of Nicaraguan society and history clashed with the world-view of most people in the countryside. In 1986 the Sandinistas tacitly acknowledged that they had made a great mistake; they reversed the state-centred agrarian reform and began to distribute land to peasant households. But by then the revolutionary fervour of the insurrection and Somoza's defeat was spent, and the difficulties of survival in the face of US opposition gave the Sandinistas little scope for assuaging peasant unrest. In the end, the Sandinistas lost the elections of 1990. Defeat came as a shock; although with the benefit of hindsight it seems clear that the FSLN's demise should have been a death foretold. Their vision of Nicaragua's past was more myth than history. While myth served the Sandinistas well in armed opposition, in power the FSLN's use of the past to guide policy-making antagonised the majority of rural Nicaraguans. More importantly, in the elections of 1990, like the elections of 2001, the US government poured in money and advisers to defeat the Sandinistas.

THEORY AND HISTORY MATTER

Debates about the extent and timing of capitalist development are not 'just academic'. The link between Sandinista history and policy is but one example of how understanding the past – trying to 'get history right' – helps us to comprehend the politics of the real world. Although it is impossible to 'get history right', as Marxists we believe that there is a past that happened, and understanding the complexities of past societies helps us to interpret the present and to think about the future. In the case of post-colonial countries, their recent histories of capitalist development often are reflected in contemporary class relations and class consciousness, which in turn have a bearing on strategies for radical social change.

The Marxist approach to history stands postmodern history writing on its head. Postmodern historians argue that subjectivity and relativity so condition all events – both how they took place and how they are understood – that there is no such thing as 'a past that happened'. History is only interpretation: ergo any one interpretation of the past is as good (or bad) as any other (Munslow 1997, pp. 1–35). Marxists, on the contrary, see writing history as a process that involves a productive tension between trying to understand a past that happened, and interpreting the past in order to grasp important dynamics of historical change. For as Marx famously said, revolutionaries try to understand the world in order to change it.

Marxists do not believe that historical conditions *determine* the possibilities and impossibilities for revolutionary change. If we have learned anything from the history of the twentieth century, it is that the relationship between history and social change is more indeterminate than determinate. Most Marxists, however, subscribe to the idea that particular historical/material contexts, including but certainly not restricted to class and property relations, condition the possibilities and impossibilities for social change. The Sandinistas knew this; one of their great strengths was that they believed that history mattered, that the past would validate their revolutionary strategy. Like many radical thinkers of the late twentieth century, the Sandinistas argued that their country had been fully capitalist for a long time. With this framework, the Sandinistas, like other revolutionary leaders in post-colonial countries, may have exaggerated the role of the proletariat in Third World countries in leading struggles for socialist change.

Thinking about the great shifts in understanding the history of capitalism raises questions about the politics of the Old and New Left. Whereas the Old Left vision of Third World countries as uniformly 'feudal' was problematic in that it obscured social changes tending in the direction of capitalism, New Left interpretations of post-colonial countries as fully capitalist seem equally problematic; they camouflaged the activities of non-capitalist classes, and tended to underestimate the weight of non-capitalist relations, which in certain times and places remained considerable in countries of the Third World.

From the vantage point of the early twenty-first century, when capitalism is aggressively turning the entire world into its own image, studying times and places where capitalism had not transformed, or fully transformed, landscape and society can provide important lessons. Historical perspectives might allow us to distinguish what is old and what is new about globalisation – capitalism in its current stage. Revitalising controversies about capitalism's rise – and fall – in Africa, Asia and Latin America could make an important contribution to the contemporary anti-capitalist debates.

With politicians, pundits and academics across the globe proclaiming that capitalism is triumphant – that we are at 'The End of History' and 'There Is No Alternative' to capitalism (TINA), it is not surprising that many people now find it difficult to imagine that capitalism can come to an end. In our era, with critics of capitalism frequently silenced in universities, the very institutions supposedly devoted to protecting freedom of thought, we are taught to forget – not to remember – that capitalism is but one historically unique way of organising society. The apparently common-sense belief that capitalism prevailed everywhere on earth in modern times has contributed to 'naturalising' capitalism: to legitimating the notion that capitalism is the natural way for human society to be organised. In these times, it is difficult to imagine that capitalism can be overthrown. By studying the history of capitalism, especially in countries of the Third World, we can remember that capitalism had a beginning, and if capitalism had a beginning it will probably also have an ending. Another world is possible.

REFERENCES AND FURTHER READING

Brenner, R. (1976) 'Agrarian Class Structure and Economic Development in Pre-Industrial Europe', *Past and Present*, February, pp. 30–75.

Brenner, R. (1977) 'The Origins of Capitalist Development: A Critique of Neo-Smithian Marxism', *New Left Review*, July–August, pp. 25–92.

Cooper, F., Isaacman, A., Mallon, F., Roseberry, W. and Stern, S. (1993) *Confronting Historical Paradigms: Peasant, Labour and the Capitalist World System in Africa and Latin America*. Madison: University of Wisconsin Press.

Dore, E. (2003) 'Debt Peonage in Granada, Nicaragua, 1870–1930: A Non-Capitalist Transition', *Hispanic American Historical Review*.

Dore, E. and Weeks, J. (1992) *The Red and the Black: The Sandinistas and the Nicaraguan Revolution*. London: Institute of Latin American Studies Research Papers.

Foster-Carter, A. (1978) 'The Modes of Production Controversy', *New Left Review*, July–August, pp. 47–77.

Frank, A. G. (1969) *Latin America: Underdevelopment or Revolution?* New York: Monthly Review Press.

Hilton, R. (ed.) (1976) *The Transition from Feudalism to Capitalism*. London: Verso.

Munslow, A. (1997) *Deconstructing History*. London: Routledge.

Polanyi, K. (1944) *The Great Transformation: The Political and Economic Origins of Our Time*. Boston: Beacon Press.

Stoler, A.L. (1995) *Capitalism and Confrontation in Sumatra's Plantation Belt, 1870–1979*. Ann Arbor: University of Michigan Press.

Wallerstein, I. (1979) *The Capitalist World-Economy*. New York: Cambridge University Press.

Weeks, J. (1985–86) 'Epochs of Capitalism and the Progressiveness of Capital's Expansion', *Science and Society*, Winter, pp. 414–36.

Weeks, J. and Dore, E. (1979) 'International Exchange and the Causes of Backwardness', *Latin American Perspectives*, 6(2), pp. 62–87.

Wheelock, J. (1979) *Imperialismo y dictadura: crisis de una formacíon social*. Mexico: Siglo Ventiuno.

Wood, E.M. (1999) *The Origin of Capitalism*. New York: Monthly Review Press.

13   Developing Country Debt and Globalisation

John Weeks

This chapter considers the relationship between developing country external debt and the integration of world markets in the 1980s and 1990s ('globalisation'). Governments, companies, other institutions, and individuals contract debt, and it is a major confusion to refer to the debt of 'countries'. Perhaps the defining characteristic of an external debt is that the contracting party is not, in general, the party upon whom the burden for repayment falls. In general, governments and the wealthy in developing countries contract the debt, while it is the mass of the non-wealthy population that bears the burden of its repayment. This is especially the case during periods of financial crisis.

DOMESTIC AND EXTERNAL DEBT

External debt has a long history, but was incidental to the world economy prior to the capitalist epoch. One of the first major external borrowings involving developing countries was by Latin American governments in the mid-nineteenth century. For the most part, the purpose of these loans was to finance infrastructure development, such as railroads and ports, in order to facilitate trade with the emerging capitalist powers. Among others, the Peruvian government defaulted on outstanding loans. The defaults did not result in a suspension of lending, for again in the early twentieth century Latin American governments sought loans, which private banks in the capitalist countries were eager to extend. During the Great Depression of the 1930s and during the Second World War there were further defaults, most notably by Mexico. These defaults did not result in suspension of lending for two major reasons. First, developed country banks served as intermediaries for loans of the 1920s, rather than as direct lenders. That is, they sold the Latin American debt on the bond market, thus incurring no risk of default themselves. Second, after the Second World War moderate to rapid

175

growth in several Latin American countries ensured the creditworthiness of the borrowing governments.

This chapter deals with developing country debt after the Second World War, and one can distinguish three periods. From 1945 until the oil crisis of 1973–74, lending to developing countries was almost exclusively to governments, from the international financial institutions, the International Monetary Fund and the World Bank. During the rest of the 1970s, the governments of underdeveloped countries in which capitalism was more highly developed, and those with considerable mineral wealth, borrowed directly from the commercial banks of the advanced capitalist countries, which gave rise to the so-called debt crisis of the early 1980s. The remainder of the 1980s was characterised by the intense efforts of the commercial banks, aided by the governments of their home countries and the international financial institutions, to recover as much of the outstanding debt as possible. During the 1980s a new phenomenon emerged, which reached major proportions in the 1990s, namely private companies and banks in developing countries borrowing directly from commercial banks in the advanced capitalist countries. This chapter focuses on this third period, which was the result of deregulation of currency markets in developing countries.

To understand why there were these three periods requires consideration of the dynamics of capital and the nature of underdevelopment. Capital is a social relation in which money is advanced with the purpose of generating more money. That is, capital is not money, commodities or means of production, but takes the form of all of these in its cycle of reproduction (see Chapter 1).

Capitalist trade among countries involves either the purchase of raw materials and intermediate products from underdeveloped countries (conversion of money capital into commodity capital), the purchase of final products from those countries for resale in the capitalist countries at a profit derived from control of supply (market power), or the sale in underdeveloped countries of commodities produced in advanced capitalist countries (conversion of commodity capital into money capital). The sale of commodities can also lead to the export of commodity capital. When companies based in advanced capitalist countries establish production facilities in underdeveloped countries, productive capital is exported. If financial capitalists make loans to governments or companies in underdeveloped countries, this involves the export of money capital. In effect, this introduces an additional step into the circuit.

Therefore, loans to governments, companies, and individuals in other countries involve the export of money capital. This export of money capital differs from loans within a country in important ways. Most generally, it involves two currencies, the currency of the lender and the currency of the borrower. The revenue of the borrower, which will be used to repay the loan, will come partly or wholly in the currency of the country in which the borrower is located. In the case of a government borrower, the revenue will derive from taxes, and for a private borrower from domestic and foreign sales. In order to repay, the domestic currency must be converted to the currency of the lender.

The 'external' character of the debt (involving at least two currencies) means that while a domestic borrower faces the problem of generating enough revenue to repay the debt, the external borrower must do this, and must also be able to convert his/her national currency into the currency of the borrower. The differences between domestic and external debt are summarised in Table 13.1. These differences arise from the apparently simple issue of conversion of the borrower's currency into the lender's. Following Marx's analysis, one sees that the problems of repayment of external debt represent an extreme example of the problems arising from money's use as means of carrying out exchange (*means of exchange*) and as means of cancellation of a debt (*means of payment*).

A debt is contracted at a given amount by a promise to pay at some future date. When the principal of the debt falls due; the value of the contracted amount may have changed. This can occur within an economy due to falling or rising prices, but the intervention of a currency exchange dramatically increases the probability that the value of the debt at the point of repayment will be different from the value when it was contracted. Formally, the difference often results from change in the exchange rate, which can be provoked by a range of causes: changes in export and import prices, capital flight from the debtor country, or the infamous 'loss of market confidence'.

The difficulties in repayment are increased if governments guarantee the external debt of the private sector. Prior to the 1980s, governments of developing countries typically restricted the convertibility of their currencies; for example, they required all foreign exchange earned by the private sector to be deposited into the central bank, and conversion to foreign currencies required government approval. In such circumstances, private sector external debt was small or non-existent, as governments gave explicit

Table 13.1 Differences between Domestic and External Debt

Type of Loan	Possible problems	Consequence
Loans within a country	Borrower fails to generate sufficient revenue to service the loan due to: – borrower-specific problems – contraction of the market for the product – general contraction of the economy	Bankruptcy, repossession of assets by lender, loss of borrower's 'creditworthiness'
Loans between countries	All the above, plus: – devaluation of the borrower's currency – inability to convert local currency into lender's currency	In addition to non-payment of government debt, if the government guarantees the private sector's debt: General financial crisis, contraction of the economy, possible fall of the government

guarantees to the lender to pay the debt should the private sector fail to do so. Once governments deregulated currency trading, guarantees of private sector external debt were no longer necessary, since companies and banks had free access to foreign exchange. In principle, free convertibility eliminates the difference between domestic and external debt, and the consequence of any non-payment by the private sector should be bankruptcy according to the rules of markets.

However, in practice, private sector failure to service external debt, even with free convertibility, led to *ex post facto* government guarantee of that debt. Perhaps the most infamous case of this occurred in Chile in the 1980s. Following sound market logic, the Chilean dictator Augusto Pinochet announced in 1982 that his government would not assume responsibility for foreign debts contracted by the private sector. However, within days pressure from banks in the United States, conveyed via the US government, forced the dictator to reverse his stand, provoking the particular form of Chile's debt crisis, namely an inability of the *government* to service the debt without a dramatic contraction of the Chilean economy. This contraction was required in order to generate trade surpluses for debt service.

In summary, governments acquire external debt for the purpose of public sector investment or to cover deficits in the balance of payments (usually trade deficits). Prior to the 1970s, governments contracted these debts with the international financial institutions. In the 1970s, external debt remained largely that of governments, contracted with private commercial banks. With the deregulation of currency markets in the 1980s and 1990s, the door opened for private sector external debt and associated financial crises.

PATTERN OF DEVELOPING COUNTRY DEBT

In the 1990s progressives vigorously took up the call to cancel developing country debt. Almost without exception, the call was for the cancellation of 'official' debt; that is, debt owed to the govern-ments of advanced capitalist countries and the international financial institutions. Those holding this position frequently based their advocacy on the following arguments: (a) that underdeveloped countries, especially the poorest, suffered from high indebtedness; and (b) that cancellation of debt would have a substantial impact on the growth potential of the indebted countries. While an argument can be made for the cancellation of the official debts of the poorest countries, both of these arguments were wrong. Most heavily indebted countries in terms of absolute debt were the middle-income countries, in Latin America and Asia, not the poorest countries. And for the poorest countries, debt cancellation would have a minor impact on growth. From the perspective of the capitalist world market, the importance of external debt lay in its close relationship to financial crisis, rather than to the debt burden itself.

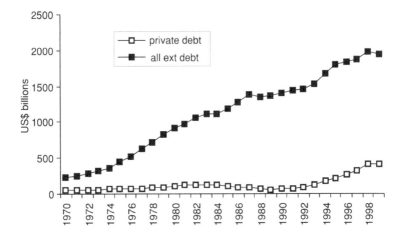

Figure 13.1   External Debt of Developing Countries, 1970–99 (constant US$ of 1995, billions)

Figure 13.1 shows the total debt and private sector debt of developing countries, and Figure 13.2 disaggregates the total debt by region. In both charts debt was divided by the US GDP deflator to adjust for inflation. These two charts demonstrate the points made above. Since 1970, total debt of underdeveloped countries has grown at a relatively constant rate, except for the 1980s, when it was virtually constant. In contrast, private sector debt was quite small until the end of the 1980s, after which it grew at an extraordinary rate of over 20 per cent per year (see Table 13.2). The only decade of rapid growth of public debt was the 1970s, when governments borrowed to cover balance-of-payments deficits that resulted from the petroleum price increases of 1973–74 and 1979. During the 1990s, when private sector debt boomed, growth of public debt was well below the rate of increase of both national income and exports; i.e. in most countries the relative burden of public debt declined.

Figure 13.2 shows that for three regions, the Middle East and North
Africa (ME&NA), South Asia (SoAsia), and Africa south of the Sahara
(SSA), the increase of total debt was quite slow after 1980. For Latin
America, debt increased in the 1980s as a result of borrowing to cover
balance-of-payments deficits, then, after holding constant for a
decade, grew rapidly in the 1990s. For East Asia and the Pacific
(which includes the South East Asian countries), there was a long
increase to the end of the 1980s, after which growth was more rapid
than before. China, with 20 per cent of the world's population,
shows a similar pattern, beginning from near zero in the early 1980s
and rising to over US$100 billion in 1995 prices.

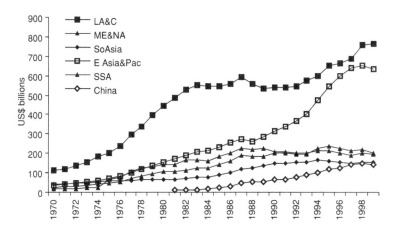

Figure 13.2   Total External Debt by Developing Region and China,
1970–99 (constant US$ of 1995, billions)

Table 13.2   Annual Rates of Growth of Public and Private Sector Debt of
Developing Countries

	Public	Private
1970–80	16.1	7.1
1980–89	5.7	−7.1
1989–99	2.1	21.6
All years	7.3	5.2

Source: World Bank, *World Development Indicators 2000*, CD Rom. These
numbers exclude the so-called transitional economies of Central and Eastern
Europe and Russia.

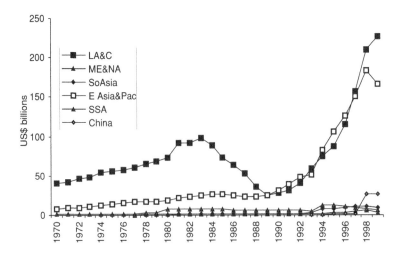

Figure 13.3   Developing Country Debt of the Private Sector, 1970–99 (constant US$ of 1995, billions)

The reason for the different growth trends is revealed in Figure 13.3, which shows private sector debt by region. For the three regions with slow growth of total debt, private sector debt is tiny. But for Latin America and East Asia, private sector debt exploded in the 1990s. This phenomenal increase, 24 per cent per year for Latin America and 21 per cent for East Asia, was the direct result of the deregulation of currency markets, which allowed private companies to borrow directly from international commercial banks. Companies were motivated to do this because borrowing costs in the advanced capitalist countries were generally lower than in the local capital markets.

Thus, with regard to external debt, one can identify three broad categories of countries. First, there are the low-income countries in which the development of capitalist production is incipient at best. These countries, most of them lying south of the Sahara, along with a few in Asia and the poorest of the Latin American countries, carry an official debt burden. With a few exceptions, this debt burden is relatively low. That many of these countries cannot service their debts reflects a general problem of national development, of which debt is a symptom rather than a cause. For another, smaller group of countries, debt burdens are low because the governments in question have not liberalised currency and capital markets, or only partially.

In this group are most of the countries of the Middle East, China, and India. Third, there are the liberalised middle-income countries of Latin America, East Asia and South East Asia. These countries, along with several of the countries in transition from central planning (not covered in this chapter), have, as a result of their governments' policies, accumulated large private sector debts.

A major difference between the Latin American countries and those of East and South East Asia is that because the latter grew so much more rapidly during 1980–97, their debt service burden declined, while that of Latin America grew in the 1990s (see Figure 13.4). However, if prior to deregulation of markets a falling debt service burden implied less vulnerability to a debt-provoked crisis, that was no longer the case in the 'globalised' 1990s. Despite a falling debt burden for most of the East and South East Asian countries, the financial crisis of 1997 struck the region with virulence. This was essentially a crisis of deregulation; since national policies of deregulation are the basis of 'globalisation', this crisis could correctly be called a crisis provoked by 'globalisation'.

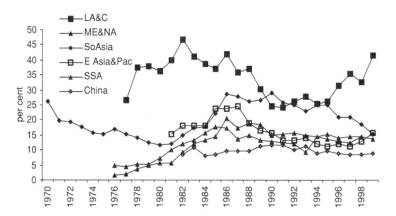

Figure 13.4   Developing Country Debt Service as a Percentage Share of Exports, 1970–99

Based on our analytical discussion and review of the pattern of developing country debt, we can provide a schematic summary of the relationship between external debt and financial crisis. Deregulation of markets creates *the possibility of a financial crisis*. In the absence of deregulation, countries would not be crisis-free, but their crises would be of a different nature. The deregulation of markets results in the accumulation of private sector external debt, so that *the form of the crisis* will be excessive debt accumulation. The proximate cause of the crisis, or 'trigger', might be a range of otherwise mundane events, such as a decline in the country's terms of trade, transitory political instability, or the perception by international currency dealers that the country would be vulnerable to a speculative attack. Once the crisis hits, the neoliberal policy orthodoxy will ensure that it passes from the financial sector to the entire economy, through imposition of high interest rates and reductions in public sector expenditure. If the government has the political independence to reject the neoliberal orthodoxy and re-regulate, the crisis may be painful, but not disastrous, as in Malaysia between 1997 and 1999. If the government zealously embraces the neoliberal policy package of austerity, the result will be disastrous (Indonesia after 1997) or even catastrophic (Argentina in 2001–02).

THE BURDEN OF REPAYMENT

In the 'globalised' 1990s financial crises leading to general economic crises resulted from debt accumulation of the private sector, such that banks and companies faced bankruptcy on a massive scale. However, these are rarely, if ever, the victims that must bear the burden of the disaster. That role is invariably reserved for the urban and rural working class, the poor peasants, and, to a lesser degree the middle class.

Indonesia between 1997 and 2001 provided an extreme, though representative, example of the relative and absolute impact of economic crisis. The speculative run on the rupiah in mid-1997 resulted in a massive and uncontrolled devaluation. Because Indonesian banks and companies had accumulated large external debts, every decline of the rupiah increased the domestic currency cost of debt service. By the time the rupiah had risen from 2,500 to over 10,000 to the dollar, the entire medium- and large-scale manufacturing sector, and the entire banking system, were bankrupt. The government's agreement to a series of IMF programmes aggravated bankruptcy. A major element in this process was the use of the domestic interest rate to attract capital from abroad and stabilise the currency. The practical effect of raising interest rates (to over 70 per cent) was to add a rising domestic debt cost to the external debt burden of the private sector.

With the private sector on the verge of total collapse, the government nationalised the entire banking sector and close to half of large-scale manufacturing. Far from using these nationalisations as a vehicle to maintain employment levels and reassert control over the financial system, the government set about a massive bailout. The collapse of the financial system implied that banks would not honour deposits, the vast majority of which were held by wealthy Indonesians. Further, loans to the manufacturing sector represented a substantial portion of the non-deposit assets of the banks. The collapse of large-scale manufacturing rendered these assets worthless. Under a so-called recapitalisation programme, the government issued public sector bonds to the nationalised banks with the purpose of entirely replacing the value of deposits and non-performing loans. The government estimated in 1999 that the bond issue would reach US$75 billion, making it, in proportion to Indonesia's national income, the largest financial bailout ever recorded. The annual interest on these bonds would by 2001 consume almost 40 per cent of the government budget and be over ten times the expenditure on health and education. To this flagrant transfer of resources to the rich was added the privatisation of the banks and manufacturing firms at 'fire sale' prices, in some cases to the pre-crisis owners. Once the banks were privatised, the interest on the bonds would accrue to their private owners. Thus, through the tax and expenditure system, there would be built into the Indonesian economy a long-run transfer of income from the poor, the working class and the middle classes to the rich. The Indonesian crisis carried a profound message: it is the institutions and dynamics of capitalist society that generate crises, and the masses of the population that bear its cost.

DEBT AND CAPITALIST INSTABILITY

Debt in and of itself is not a problem for governments. Just as private corporations borrow to finance investments, so a government may borrow to foster modernisation and development. It becomes a problem in the context of the circuit of capital and the institutional arrangements that regulate capital. The international debt crisis of the 1980s resulted from the accumulation of public debt, compounded by the shifting of the burden of private sector debt repayment to governments.

In the 1990s the deregulation of money capital flows by governments throughout the world resulted in a rapid accumulation of debt held by private companies and banks in underdeveloped countries. This debt accumulation created the possibility of crises considerably more severe than those of the 1980s, realised in South East and East Asia in 1997–98, and Argentina in 2001–02. The greater severity resulted from the adoption by governments of unrestricted convertibility of domestic currencies into foreign currencies. In an important sense, the growing private sector debt was but the tip of the iceberg of potential instability. If currencies can be converted without restriction, then the entire money supply of a country becomes 'external debt', in that it can be converted at will and sent abroad as capital flight.

Unrestricted convertibility creates an international financial market continuously on the verge of a speculative dementia, holding out the promise to capital of unlimited profit without engaging in the time-consuming process of production. Marx wrote of the capitalist pipe dream of profits without the annoyance of marshalling, supervising and disciplining workers, and without the need to satisfy the demands of consumers. The realisation of that dream is the systemic instability of capitalism.

FURTHER READING

Bitterman, H.J. (1973) *The Refunding of International Debt*. Durham: Duke University Press.

De Pinies, J. (1989) 'Debt Sustainability and Overadjustment', *World Development* 17(1).

Maddison, A. (2001) *The World Economy: A Millennial Perspective*. Paris: OECD.

UNDP (1999) *Debt and Sustainable Human Development*, Technical Advisory Paper, No. 4. Management Development and Governance Division, Bureau for Development Policy.

Weeks, J. (ed.) (1989) *Debt Disaster: Banks, Governments and Multilaterals Face the Crisis*. New York: New York University Press.

14 Globalisation and the Subsumption of the Soviet Mode of Production under Capital

Simon Clarke

Mikhail Gorbachev was elected General Secretary of the Communist Party of the Soviet Union in 1985 with the mission to reform an economic system which had progressively lost its dynamism during the 'years of stagnation'. Gorbachev's programme of 'perestroika' aimed to introduce market elements into the Soviet administrative-command system in order to subject Soviet enterprises to the discipline of world-market prices. The transition to a market economy was completed under Yeltsin, who freed most wages and prices from government control at the end of 1991.

The neoliberal Russian and Western economists who were the ideologues of Yeltsin's programme of radical reform expected that the abandonment of administrative methods and the transition to a market economy would lead to the rapid transformation of the Soviet Union into a capitalist economy as investors took advantage of the highly skilled labour force and advanced science and technology that had built up the Soviet military machine. In fact the outcome was a disaster: the longest and deepest recession in recorded human history, including a decline in industrial production twice as deep as that provoked by Hitler's invasion of the Soviet Union, and living standards which fell back to the level of the 1960s, when Khrushchev was dismissed for his economic failures.

The most common explanations for this disaster refer to the adoption of inappropriate policies by the Russian government. While the government's domestic critics argue that the collapse has been the result of the adoption of neoliberal reforms, neoliberals argue to the contrary, that the collapse has occurred because reforms have not been sufficiently radical. However, what has happened in Russia has not been the result of policy choices. The ability of policy-makers to mould the economy is constrained by the instruments at their disposal and the structural characteristics of the economy which they seek to manipulate. What has happened in Russia has its roots in the Soviet period, a result of the unfolding of the contradictions of the Soviet economic system in the context of its integration into the global capitalist economy.

THEORISING TRANSITION: SMITH AND MARX

Many commentators have compared the Soviet system to that of feudalism in being based on the appropriation of a surplus by the exercise of political power. For Adam Smith and Friedrich Hayek the central feature of feudalism was the distortion of the natural order of the market economy by the superimposition of political rule, and the transition from feudalism to capitalism depended on sweeping away the political institutions of the old regime in order to establish the freedom and security of property – what Smith referred to as 'order and good government' – which would allow the market economy to flourish (Clarke 1988, ch. 1). This was the ideology that informed the liberal project of the transition to a capitalist market economy in the former state socialist economies. According to this model the transition is not theorised as an evolutionary development of the existing system under the impact of its integration into the structures of the world market. For this model the existing system has no dynamic of its own. It is defined purely negatively as a barrier to change which must be destroyed, so that a new system can be created out of the fragments set free by its destruction. It is not to Adam Smith or Friedrich Hayek that we should look to understand the development of capitalism, but to Smith's most cogent critic, Karl Marx.

For Marx the development of capitalism was not the realisation of individual reason but an expression of the contradictions of the feudal mode of production, as the development of the forces of production broke the fetters of feudal production relations with the development of commodity production; this was massively accelerated by the dispossession of the mass of the rural population, who became the wage labourers for capital and the consumers of the products of capitalist production.[1] The dispossession of the rural population provided an ample reserve of cheap wage labour which could be profitably employed by the capitals accumulated through trade and plunder. At this first stage of capitalist development, however, capitalists did not change the methods of production which they had inherited, so the subsumption of labour under capital was purely formal. Merchant capitalists made huge profits by exploiting their commercial monopolies. Capitalist producers cut their costs not by transforming methods of production but by forcing down wages and extending the working day.

Capital only penetrated the sphere of production when competition between capitalists induced and compelled them to revolutionise the methods of production in order to earn an additional profit, or resist the competition of those who had already done so. It was only with the 'real subsumption' of labour under capital that the characteristic dynamic of the capitalist mode of production got under way. Nevertheless, in the peripheral regions of the emerging global capitalist economy the subsumption of production under capital remained purely formal, based on the intensified exploitation of pre-capitalist social forms, with the 'second serfdom' in Eastern Europe and the reinforcement of slavery and quasi-feudal forms of exploitation in the colonial world.

The process described by Marx as that of 'primitive accumulation' (see Chapter 8) was largely achieved in Russia in the Soviet period, when the peasants were dispossessed and transformed into wage labourers, not for capital but for the state. The Soviet state launched a programme of industrialisation, based on the introduction of the most advanced capitalist technology, but the social form of the production and appropriation of a surplus in the Soviet system was quite different from that characteristic of the capitalist mode of production, and the dynamics of the system were correspondingly different.

THE CONTRADICTIONS OF THE SOVIET MODE OF PRODUCTION

The Soviet system was not based on the maximisation of profit by the production of commodities, but nor was it based on the free association of producers and planned provision for social need. It was an essentially non-monetary system of surplus appropriation subordinated to the material needs of the state and, above all, of its military apparatus. The development of the system was not subordinated to the expansion of the gross or net product in the abstract, an abstraction which can only be expressed in a monetary form, but to expanding the production of specific materials and equipment – tanks, guns, aircraft, explosives, missiles – and to supporting the huge military machine.

The system of 'central planning' was developed in Stalin's industrialisation drive of the 1930s in a framework of generalised shortage.

The system was driven by the demands of the state for a growing physical surplus, with scant regard for the material constraints on production of skills, resources and capacities. The strategic demands of the five-year plan would be determined by the priorities of the regime, and ultimately by the demands of the military apparatus, which would then be converted into requirements for all the various branches of production. These requirements came to be determined in a process of negotiation between the central planning authorities, ministries and industrial enterprises.

Soviet social relations of production were supposed to overcome the contradictions inherent in the capitalist mode of production in being based on the centralised control of the planned distribution and redistribution of productive resources. However, the Soviet system was marked by its own contradiction inherent in the subordination of the system of production to the Soviet system of surplus appropriation. As in the case of feudalism, this contradiction was expressed in the development of market relations within the Soviet system which provided the basis for the emergence of new, proto-capitalist forms of surplus appropriation.

The fundamental contradiction of the Soviet system lay in the separation between the production and appropriation of the surplus. The centralised control and allocation of the surplus product in the hands of an unproductive ruling stratum meant that the producers had an interest not in maximising but in minimising the surplus that they produced. Since neither the worker, nor the enterprise, nor even the ministry, had any rights to the surplus produced, they could only reliably expand the resources at their disposal by inflating their production costs, and could only protect themselves from the exactions of the ruling stratum by concealing their productive potential. Resistance to the demands of the military-state-Party apparatus for an expanding surplus product ran through the system from top to bottom and was impervious to all attempts at bureaucratic reform. The resulting rigidities of the system determined its extensive form of development, the expansion of the surplus depending on the mobilisation of additional resources. When the reserves, particularly of labour, had been exhausted, the rate of growth of production and of surplus appropriation slowed down (Clarke et al. 1993, ch. 1).

MARKET ELEMENTS IN THE SOVIET SYSTEM

Market relations played an increasing role in the Soviet system. As in the case of feudalism, the contradictions inherent in the Soviet system meant that money, the market and quasi-market relations developed spontaneously out of attempts to overcome the contradictions of the system and were tolerated, however reluctantly, by the authorities.

First, although social reproduction was as far as possible subordinated to the imperatives of production, with a wide range of goods and services being provided through the workplace, labour power was partially commodified and workers were paid a money wage. Money in the hands of workers lubricated the black market for consumer goods and for the private production of agricultural produce for the market, which was tolerated and even encouraged; rural producers were allowed to sell their own products on the *kolkhoz* markets, which provided a basis for more extensive market transactions.

Second, Trotsky's early attempts at the 'militarisation of labour' were unsuccessful and, although wages were regulated centrally, workers were always in practice free to change jobs in search of higher wages. Labour shortages put increasing pressure on the centralised regulation of wages as employers sought to attract the scarcest categories of labour, so that wage setting had to take account of labour market conditions.

Third, while the centre could allocate rights to supplies, it could not ensure that those supplies were delivered to the right place, at the right time, and were of the desired quality, so that enterprises used informal personal connections with their suppliers, often backed up by local Party *apparatchiki*, to secure their supplies, and came increasingly to draw on the services of unofficial intermediaries, the so-called *tolchaki* (pushers), who were the pioneers of market relations within the Soviet economy. The central directives which nominally regulated inter-enterprise transactions within the Soviet system were therefore only realised in practice through exchanges within networks of personal, political and commercial connections which provided the basis for the emergence of financial and commercial intermediaries under perestroika.

Fourth, the need to acquire advanced means of production from the West meant that the Soviet Union had to export its natural resources in order to finance its essential imports of machinery. The 1930s industrialisation drive was made possible by the massive export of grain forcibly expropriated from the peasantry, which led to the devastating famines of the 1930s. By the Brezhnev period the Soviet Union had become dependent on its exports of oil and gas to finance its imports of machinery and even of food, and the reproduction of the Soviet system depended increasingly on trans-actions on the world market. In 1985 fuel accounted for more than half the Soviet Union's exports, with another quarter being accounted for by raw and semi-processed raw materials, while machinery accounted for a third of imports and food for a fifth. The share of world trade in the net material product of the Soviet Union increased from 3.7 per cent in 1970 to a high of 11 per cent in 1985, while oil and gas production doubled between 1970 and 1980. At the same time, the Soviet Union saw a sharp improvement in its terms of trade, primarily due to rising fuel prices, the net barter terms of trade improving by an average of 5 per cent per annum over the period 1976–80, and 3 per cent per annum between 1980 and 1985 (IMF/World Bank/OECD 1991, vol. 1, pp. 86, 105), helping to offset the decline in productivity growth and allowing the Soviet Union to increase its import volume by a third, while export volume increased by only 10 per cent. The improved terms of trade also made a substantial contribution to the buoyancy of government revenues through the price equalisation system, according to which the state appropriated the difference between domestic and world market prices. This opening of the Soviet economy to the world market, and the corresponding political processes of détente, were by no means a sign of fundamental change in the Soviet system, but were rather the means by which change was constantly postponed. However, such favourable circumstances could not last: production of gas and oil peaked in 1980, so that the Soviet Union was increasingly dependent on improvement in the terms of trade to sustain its economy. When the terms of trade turned sharply against the Soviet Union from 1985, reforms could be postponed no longer.

THE TRANSITION TO A MARKET ECONOMY

The 'transition to a market economy' was not an alien project imposed on the Soviet system by liberal economists, but in the first instance was an expression of the fundamental contradiction of the Soviet system. The first stage of market reforms sought to improve the balance of external trade by ending the state monopoly of foreign trade and licensing enterprises and organisations to engage in export operations and to retain a portion of the hard currency earned. The idea was that this would give industrial enterprises an incentive to compete in world markets and to use the foreign exchange earned to acquire modern equipment. In practice it provided a windfall for exporting enterprises, at the expense of the state, and opportunities for those with the right connections to make huge profits by acting as intermediaries.

Once the precedent had been set, other enterprises sought the right to sell above-plan output on export or domestic markets, and to retain a growing proportion of the proceeds. Allowing enterprises to sell on the market provided an alternative source of supply to the centralised allocations which the state could not guarantee, and if the state could not guarantee supplies, why should enterprises continue to deliver their state orders when they could sell more profitably at market prices? Thus the development of market relations undermined the control of the centre, created a space for the development of capitalist commercial and financial enterprise and precipitated the collapse of the administrative-command system. Rather than resolving the contradictions inherent in the Soviet system, as Gorbachev had hoped, the transition to a market economy brought those contradictions to a head. The surplus which had been appropriated by the state was now retained by enterprises or appropriated by the new financial and commercial intermediaries which arose to handle the emerging market relations.

Yeltsin's programme of radical reform was no more than a belated recognition of the fact that the state had lost control of the management of the economy. The decision to free wages and prices from state control was a recognition that the state had already lost control of wages and prices, since by the end of 1991 nothing was available to buy at state prices. Corporatisation and privatisation of state enterprises was an equally inevitable consequence of the development of a market economy, merely a juridical recognition of what had already become a fact: that these enterprises had already detached themselves from the administrative-command system of management which no longer had any levers of control over them. Privatisation did not give enterprises any more rights than they already had, while it allowed the state to abdicate all the responsibilities to them which it no longer had the means to fulfil. Thus, the rhetoric of neoliberalism and radical reform was little more than an ideological cover for what was essentially a bowing to the inevitable.

RUSSIA'S CAPITALIST TRANSITION: NEW FORMS OF SURPLUS APPROPRIATION

The surplus appropriated by the Soviet ruling stratum took the form of the material goods which sustained Soviet military might and the lifestyle of its ruling stratum, but these goods were produced at enormous cost. According to the calculations of Western economists, at world-market prices a large proportion of the Soviet economy was 'value-subtracting', primarily because of the very high energy and raw material intensity of production. In value terms, the bulk of the surplus was accounted for by the rents appropriated through the export of fuels and raw materials. The result of the subordination of the Russian economy to global capitalism was, therefore, inevitable: massive profits would accrue to those who controlled the export of fuels and raw materials, while the bulk of domestic industry and agriculture would not even be able to cover their input, let alone their wage costs. The only policy issues were those of the extent to which the state would be able and willing to divert the profits of the exporters to subsidise loss-making domestic producers and support domestic investment in new technologies.

The development of a market economy in Russia and the emergence of private commercial and financial capitalist enterprises represented a change in the form of surplus appropriation, the surplus being appropriated in monetary rather than in material form. The new capitals were formed out of the commercial and financial intermediaries which had been rooted in the Soviet system and been given free rein by perestroika. They appropriated their profits by establishing the monopoly control of supplies which had formerly been the prerogative of the state. They acquired this control on the basis of rights assigned to them by state bodies, including property rights acquired on the basis of the privatisation of state enterprises, and the privatisation of the banking system, and they maintained their control, where necessary, by the corruption of state officials and enterprise directors, backed up by the threat and use of force. However, the change in the form of surplus appropriation was not matched by any change in the social relations of production (Clarke 1996).

The surplus was not appropriated on the basis of the transformation of the social organisation of production or the investment of capital in production. Investment declined steadily, to less than a quarter of its 1990 value in 1998. The average age of industrial plant and equipment in the Soviet period was about nine years, but by 1999 it had increased to over 18 years, with less than 4 per cent being less than five years old and about two-thirds having been installed before the beginning of perestroika (all data in this section is from Goskomstat 2000 and Goskomstat 2001). Far from being regenerated by the transition to a market economy, the productive economy was still capitalising on the deteriorating legacy of the past.

The surplus was appropriated by the notorious oligarchs, who have privatised the former state monopolies of banking and foreign trade, siphoning off enormous profits, estimated at US$20–25 billion per year, which are transferred abroad to offshore accounts. The bulk of the profits of the oligarchs derive from the sale of Russian fuel and raw and processed raw materials, which make up 80 per cent of Russian exports, on world markets, but they make almost no investment even in the oil and gas and metallurgical companies which supply them, so that the production of fuels is declining, existing reserves are rapidly being depleted and the exploitation of new reserves is postponed because of the lack of investment. Oil extraction fell by 42 per cent between 1990 and 1999. In 1999 the rate of fixed investment as a proportion of output in the oil industry was less than a fifth of the 1985 level. Even gas production fell by 15 per cent between 1990 and 1999, but labour productivity fell by more than half, while investment as a proportion of output had fallen by 40 per cent since 1985. The privatisation of export revenues led to a massive fall in federal government revenues and growing reliance on debt finance. Although the oligarchs were induced to pay substantial amounts to the federal government in tax and royalty payments, they recovered some of this through the banking system, through speculation and investment in government debt, debt service now amounting to 3.6 per cent of GDP.

While the oligarchs have privatised much of the surplus that was formerly appropriated by the state, the possibilities of profiting directly from industrial investment are minimal. The windfall profits which enterprises could make in the late 1980s, when they could buy at state prices and sell at market prices, were annihilated by the liberalisation of prices at the end of 1991. With the collapse of the Soviet system, enterprises inherited their premises, capital stock and stocks of parts and raw materials, which enabled many to remain in profit by trading on their inherited assets; but by 1996 the majority of enterprises were loss-making, the figure only falling to 41 per cent in the recovery of 1999. The bulk of the remaining enterprise profits are annihilated by taxation, leaving little or nothing to pay out as dividends to shareholders. While the taxation of enterprise profits amounted to 4.9 per cent of GDP in 1999, dividends amounted to only 0.5 per cent of GDP, up from 0.3 per cent in 1998. The traditional state enterprises, the majority of which have been privatised, have struggled to survive by any means that they can with the limited resources at their disposal: seeking out new markets, deferring payments to the government, their suppliers and their employees, looking for subsidies from local and federal government, and looking for profitable connections with criminal organisations or foreign companies; but industrial production halved between 1990 and 1999, with the production of light industry falling by 85 per cent as imports flooded the domestic market.

Meanwhile, new capitalist enterprises are concentrated in trade, catering and services, with much less penetration of construction, transport and communications and minimal penetration of industry and agriculture. New capitalist enterprises are mostly small unincorporated private companies, paying low wages and making small profits. In October 1999, average wages in the private companies which dominate trade and catering were only two-thirds of the wages paid in the remaining state enterprises, half the wages paid by incorporated companies and a fifth of the wages paid by foreign companies. Low wages, however, were not associated with high profits: almost half the companies in trade and catering were loss-making in 1998.

RUSSIA IN THE GLOBAL ECONOMY: COMPARATIVE PERSPECTIVES

I have focused on the internal dynamics of the transition to a market economy, but it should be apparent that the transition has been driven by internal forces unleashed by the integration of the Soviet system into global capitalism as a classic neo-colony, producing cheap fuels and raw materials for global capitalism and importing foodstuffs and manufactured goods while domestic production languished, unable to compete with its archaic production technology and inappropriate social organisation of production in the face of unfavourable market conditions. As in the classic case of neocolonialism, the surplus is appropriated by multinational corporations and their comprador capitalist partners. Foreign direct investment between 1994 and 1999 amounted to only US$3 billion per annum. In 1999, 23 per cent of foreign investment went into oil and metallurgy, 20 per cent into trade and catering, commerce and finance, and 15 per cent into the food processing industry, with only a trivial amount in the remaining industrial branches.

Meanwhile, the subsumption of labour under capital within Russia remains overwhelmingly purely formal. The vast majority of

Russian enterprises struggle to survive in the face of intense domestic and foreign competition, with minimal investment and earning little or no profits, using inherited plant and equipment and retaining the traditional Soviet social organisation of production, while the bulk of any surplus they produce is appropriated by monopolistic and at best semi-criminal commercial and financial intermediaries. Enterprises cut costs not by revolutionising production methods, but by reducing real wages and intensifying labour; and they stay in business by defaulting on their payments to suppliers, the government and to their own employees.

The fate of Russia has not been determined exclusively by its own historical legacy. While the other Soviet republics, as well as Romania and Bulgaria, have suffered from the collapse of the Soviet system as badly as Russia has, most of the former Soviet satellites in Eastern Europe soon recovered from the transition crisis; and the experience of China, of course, presents almost a mirror image of the fate of Russia in its transition to a market economy. While Russian GDP per head fell by almost a half over the 1990s, in China it doubled. While industrial production in Russia fell by more than half, in China it increased more than three times. While agricultural production in Russia fell by almost a half, in China it increased by 50 per cent.

Many commentators attribute these differences to the different policies pursued by the various national governments. The international financial institutions have contrasted the fate of Russia at various times with the success of Poland, Hungary and the Czech Republic. Critics of neoliberalism contrast the fate of Russia with the success of China. But policy-makers in all these countries have been severely constrained by the circumstances in which they have found themselves and restricted by the opportunities they confront. Russia has not pursued radically different policies from those of her former satellites, while much of the programme of perestroika was similar to the reforms being introduced at the same time in China. It is not so much the policy packages which have differed, all of which have been based on the subordination of the domestic economy to world market prices, as the outcomes.

In all of these countries the 'transition to a market economy' has not been so much a feature of a particular set of policies, as a strategy of integration into global capitalism. The specificity of Russia lies not in the policies pursued by its government but in the mode of its integration into global capitalism, which has been dictated by the dynamics of the latter. Against many of the other former Soviet republics, Russia at least had the advantage of having stupendous natural resources, in the form of oil, gas, metals and minerals. But the former Eastern European satellites had the advantages of a highly skilled and relatively low-paid industrial labour force and of location on the fringes of the European Union, giving them ready access to a booming market and making the economies very attractive to foreign investors. China, on the other hand, had the advantages of location on the Pacific rim, of political stability and, above all, of abundant reserves of cheap labour. The latter allowed China to pursue a dualistic strategy of continuing to subsidise the strategically important traditional state industries, while encouraging private and foreign investment in local and foreign-owned enterprises. Nevertheless, the unevenness of the Chinese pattern of development creates its own problems, which raises the question of how long such a dualistic strategy can be sustained. But that is another question.

REFERENCES AND FURTHER READING

Clarke, S. (1988) *Keynesianism, Monetarism and the Crisis of the State.* Cheltenham: Edward Elgar.
Clarke, S. (1996) 'The Enterprise in the Era of Transition', in S. Clarke (ed.) *The Russian Enterprise in Transition.* Cheltenham: Edward Elgar.
Clarke, S., Fairbrother, P., Burawoy, M. and Krotov, P. (1993) *What about the Workers? Workers and the Transition to Capitalism in Russia.* London: Verso.
Goskomstat (2000) *Rossiiskii Statisticheskii Ezhegodnik.* Moscow: Goskomstat Rossii.
Goskomstat (2001) *Rossiya v Tsifrakh.* Moscow: Goskomstat Rossii.
IMF/World Bank/OECD (1991) *A Study of the Soviet Economy.* Washington/ Paris: IMF, World Bank, OECD.

NOTE

1. This was, of course, not an automatic process but one that was mediated through the class struggle, as Robert Brenner classically argued; but the class struggle itself expresses the dynamics of the contradiction between the forces and relations of production.

Part III

Crisis and the Supercession of Capitalism

15 Capital Accumulation and Crisis

Paul Zarembka

'Crisis' can be used and misused. If it is used too frequently and for circumstances not clearly defined, it loses meaning. It can become an escape from deeper understanding. Virtually everyone would agree that the 1930s represented a world economic 'crisis' and one of major proportions. It led to the success of fascism and the resulting world war. In China, it eventually led to renewed civil war and revolutionary success. But what of lesser examples, such as the so-called 'oil crisis' of the early 1970s? Virtually everyone would agree that Argentina is in crisis in 2002. But what of other South American countries at the same time?

Crisis in capitalism has economic and political dimensions and always includes the extent to which workers are mobilising. Short-term or conjunctural crises of capitalism have multiple causes and the most we can expect of theory at present is to understand a pre-disposition to crisis. Theory for the secular crisis of capitalism, however, is available within the Marxist tradition, although not to be found directly in Marx. That is, in subjecting the concept of 'accumulation of capital' to a deeper understanding and using the work of the Polish-German revolutionary leader Rosa Luxemburg as a stepping stone, limitations in Marx can be overcome and the secular crisis of capitalism understood. In the process, along with the silent compulsion of the market in the struggle between capital and wage labour over surplus value, it is necessary to incorporate the roles of force and violence integrally into understanding accumulation and crisis.

MARX ON CRISIS

Marx does not develop an explicit crisis theory. But he does clearly indicate what to examine regarding the possibility of crisis. Such possibility arises from the fact that the mere production of a commodity does not guarantee sale; rather, sale is limited 'by the proportional relation of the various branches of production and the consumer power of society' (Marx 1894, p. 244). The former problem, of disproportionalities, is well accepted even by many non-Marxists. That is, within capitalism there is no guarantee that one sector (branch) of the economy actually produces exactly what other sectors need from that sector, particularly in circumstances of the constantly changing technologies being developed within capitalism. This theory was emphasised by the Ukrainian economist Tugan-Baranowsky and is well presented by Hilferding (1910, Part IV) whose work was an important background for Bukharin's and Lenin's works on imperialism.[1]

The role of overall consumer spending power is more controversial. Can production exceed the ability to pay, can there be 'overproduction' for society as a whole? Marx devotes considerable attention to the views of Ricardo on the dynamics of capitalism and in the process lays out his approach to understanding this overproduction possibility of crisis (Marx 1905, pp. 492–535). He says that Ricardo's erroneous conception denying the possibility of overproduction is based upon seeing products as being exchanged against products, rather than understanding capitalist production as concerned with the expansion of surplus value (pp. 493–5). The first elements of capitalist production are 'the existence of the product as a commodity, the duplication of the commodity in commodity and money, the consequent separation which takes place in the exchange of commodities and finally the relation of money or commodities to wage-labour' (p. 502). The capitalist first wants to turn commodity capital back into *money capital*. While forced sales (selling mandated by a need to pay) are often an important element of crises, in any case, sales are necessarily limited by those needs which are backed by ability to pay. Therefore,

> *overproduction* is specifically conditioned by the general law of the production of capital: to produce to the limit set by the productive forces, that is to say, to exploit the maximum amount of labour with the given amount of capital, without any consideration for the actual limits of the market or the needs backed by the ability to pay; and this is carried out through continuous expansion of reproduction and accumulation, and therefore constant reconversion of revenue into capital, while on the other hand, the mass of producers remain tied to the average level of needs, and must remain tied to it according to the nature of capitalist production. (Marx 1905, pp. 534–5)

Marx later notes that the features of bourgeois distribution 'enter into bourgeois production itself, as a determining factor, which overlaps and dominates production' and this, in turn, is the deepest cause of crises (Marx 1910, p. 84).

A concise, but very readable, aid in understanding Marx's approach to crisis is provided by Kenway (1987). Clarke's (1994) full-length study concludes that Marx did not actually have a theory of crisis as such, but that he demonstrated the important proposition that 'the tendency to crisis is pervasive ... as the tendency to over-production runs into the barrier of the limited market' (p. 279). By the time *Capital* was written, he says, Marx was more interested in the secular development of capitalism than in a conjunctural crisis theory. Of course, the fact that Marx himself did not have an explicit theory of crisis does not mean that such a theory cannot be built on Marxist foundations.

CRISIS FROM UNDERCONSUMPTION OR FALLING PROFIT RATES?

Underconsumption?

Underconsumption theories of crisis emphasise insufficient effective demand for consumer goods, principally because the level of worker wages is unable to sustain sufficient demand relative to production levels. The level of wages in different countries, and in different sections and sectors of a particular country, is a result of class struggles and historical developments, with capital continually striving for lowering of wages. This capitalist pressure for lowered wages threatens an underconsumption (relative to production). Some have thought that a theory of 'increasing misery' for workers could be taken from Marx, but Lapides' careful research (1998) has dispelled this notion. Desai (1991) provides an overview of the case for, and the limitations of, underconsumption theory, while Bleaney (1976) undertakes the detailed analysis.

Certainly, there is a relationship between underconsumption and conceptualising overproduction, with underconsumption more narrow in focus as a source of crisis in capitalism. Overproduction refers to all sources of supply of commodities exceeding their demands, whether for consumption goods for workers or capitalists, for produced means of production, for other social classes within the existing capitalist structure, or for those classes existing within a pen-etration of non-capitalist structures. Marx himself virtually never referred to underconsumption.

Falling Rate of Profit?

There may be an understanding of crisis centred around a tendency of the rate of profit to fall. But, then again, there may not be. Here is the problem. The rate of profit is the measure capitalists use to express their 'return' on their investment in means of production. Since the basis of profit, interest, and rent in a capitalist society is surplus value s, and surplus value is measured in terms of hours of work not returned to those who do the production (the workers), the rate of profit r (abstracting, for this discussion, from interest and rent) is the ratio of surplus value to the socially necessary labour time going into the means of production being used by workers, C; i.e. the rate of profit $r = s/C$. On the other hand, the rate of surplus value is s/v, where v is the variable capital of the capitalist – the cost of labour power to the capitalist measured in terms of value. The rate of profit can therefore be re-expressed as $r = s/v \div C/v$, or after a little algebra to clearly indicate the role of s/v, as

$$ r = \frac{s/v}{\left(\dfrac{s}{v} + 1\right)\left(\dfrac{C}{s+v}\right)} $$

$C/(s+v)$ describes the ratio of the value invested by capital in means of production relative to the labour power currently provided by workers. The rate of profit is falling, suggesting the possibility of crisis, if this ratio $C/(s+v)$ is rising while the rate of surplus value s/v remains fixed (or rises only slowly).[2]

From casual observation, it is easy to accept that wage labourers today are working with more massive means of production than 50, 100 or 200 years ago. But this is not good enough. For, we are discussing the *socially necessary labour time* which goes into the means of production and there are certainly technological improvements in the very making of means of production, thus reducing those labour requirements over time. Furthermore, the production of relative surplus value discussed by Marx is precisely targeted toward increasing the rate of surplus value, s/v. Therefore, how could we assert a falling tendency of the rate of profit? The manner in which Marx does it in Volume 3 of *Capital* is initially to take the rate of surplus value, s/v, as fixed and to abstract from technological improvements in producing means of production so that $C/(s+v)$ is 'clearly' rising.[3] Ergo, he gets a falling tendency for the rate of profit. When he removes these assumptions Marx describes countervailing factors as 'counter-tendencies'. This is all well and good, but it does not show whether capitalism really is described theoretically by some type of law of a tendency for the rate of profit to fall. And, while falling profits in the first half of the nineteenth century encouraged economists to see this as a fact requiring theoretical explanation, a century and a half later of rising, falling, rising, falling (etc.) rates of profit pushes into the background even searching theoretically for a tendency in profit rates.

As Clarke (1994, pp. 58–72) points out, a falling tendency in the rate of profit as a backbone toward understanding crises did not arise within Marxism until the 1970s, and the resulting discussion is not very convincing.

TECHNOLOGICAL CHANGE AS POTENTIALITY FOR CRISIS

As an element conditioning the possibility of crisis, technological change in capitalism derives from the capitalist struggle to increase that portion of the workers' labouring day which is withheld from workers. Workers are only employed by capital insofar as the time they work is greater than the time required to produce (via other workers) the necessities of all workers for maintaining life and strength for further work, both for themselves and for their children. Those necessities are both biological and social/historical in a very complicated process. If those necessities can be produced with less expenditure of labour time (including time required to produce means of production), this leads to capital getting more from the same workday. This result Marx calls production of relative surplus value (see Chapters 1 and 5).

Technical change is focused on reducing the time required to produce such items as clothing (the 'Industrial Revolution') and food (the 'Green Revolution'). Marx labelled the capitalist establishments in industries producing these goods as being within 'Department 2'. Those engaged in producing means of production are in 'Department 1', the department producing the instrumentality of capitalist control. Of course, technical change occurs in Department 1 also; iron is revolutionised into steel.

Technical change has contradictory implications and thus does lead to possibilities for crises, including disproportionalities among sectors as well as crises of overproduction. It may reduce the required labour power for production but, in so doing, it also decreases values being produced since values refer to labour time. And it can throw more commodities on the market, requiring more outlets. Yet, capital in general, in accumulating, strives to exploit more labour power with concomitant requirements for more means of production, which in turn requires more market outlets for commodities.

CAPITAL ACCUMULATION

Capitalism is a social system driven by capital accumulation and so accumulation must be clearly defined. Were it to mean more means of production, such as more equipment in factories and on agricultural lands, it would mean the same as in mainstream economics. Yet for Marx capitalism is fundamentally a social relation between capital and wage labour, between capitalists and workers working for a wage. It is a relation of exploitation and of power, derived from one social class controlling the main means of production. Accumulation of capital, therefore, needs to be seen as the extension of that social relationship, i.e., as incorporation of more wage labourers under the domination of capital, with the concomitant requirement for more means of production.

Marx was insufficiently clear in utilising the concept 'accumulation of capital'. The norm is given by clear statements such as 'accumulation reproduces the capital-relation on a progressive scale, more capitalists or larger capitalists at this pole, more wage-labourers at that' or 'capital is not a thing, but a social relation between persons, established by the instrumentality of things' (1867, pp. 575, 717, the latter commenting favourably on Wakefield).[4] Nevertheless, sometimes he can be read as meaning more means of production. This ambiguity has led to problems in the theory of capitalism. Indeed, after the success of the Bolshevik revolution in 1917, undeserved priority was given to Lenin's understanding of Marx's economics which included understanding accumulation of capital as increased production (see Zarembka 2000).

Luxemburg (1913) undertook the most penetrating analysis of accumulation of capital and advanced on Marx's theory. Her *Accumulation of Capital* of 450 pages represents one of the longest and most comprehensive works in all of Marxist economic theory, outside of Marx's own work.

In *Capital* Marx characterised the economy as being only capitalist, with no other social classes than capitalists and workers (and landlords, in some places). While he was quite aware of the existence of other classes, this delimitation, says Luxemburg, got Marx into trouble when analysing the accumulation of capital. If the economy is assumed to be only capitalist, then 'the desire to accumulate plus the technical prerequisites of accumulation is not enough ... to ensure that accumulation can in fact proceed and production expand: the effective demand for commodities must also increase. Where is this continually increasing demand to come from?' (p. 131). Capitalists sell to workers the monetary equivalent of their subsistence needs, but oppose anything further; indeed, workers buying consumer goods 'merely refund to the capitalist class the amount of the wages they have received, their assignment to the extent of the variable capital' (p. 132). Capitalists also sell to themselves subsistence and luxury goods, but the drive within capitalism is quite distinctly for *accumulation*, not merely luxury consumption. So, the capitalists' only other outlet is marketing means of production.

Yet, the system must reach an impasse since, within the capital and wage-labour relation, we cannot answer for whom the additional means of production would be produced. 'From the capitalist point of view it is absurd to produce more consumer goods merely in order to maintain more workers, and to turn out more means of production merely to keep this surplus of workers occupied' (p. 132). Thus, the constant creation of a home market or the advance of imperialism into areas not yet, or not fully, capitalist is necessary: 'the decisive fact is that the surplus value cannot be realised by sale either to workers or to capitalists, but only if it is sold to such social organisations or strata whose own mode of production is not capitalistic' (pp. 351–2). In other words, capital must reach beyond itself.

Luxemburg's *Accumulation of Capital* is difficult reading because it is so rich in understanding. A decisive implication is the need to move away from 'purely' economic, market, issues and also to consider the penetration of non-capitalist regions. She undertakes this in the last portion of her book.

Marx had commented that 'constant capital is never produced for its own sake but solely because more of it is needed in spheres of production whose products go into individual consumption' (Marx, 1894, p. 305).[5] While quite conscious of capitalism's drive for production and the accumulation of capital for its own sake, Luxemburg confronted that drive with an additional reality, the reality that the commodities to be produced by workers with the aid of means of production must ultimately find a target in consumption (producing a new railroad track adjacent to an old one simply to provide rail traffic for the old track consisting of construction materials for that new track hardly cuts ice in defending a proposition of capitalism's freedom for continuous expansion in means of production).

Although failing to clarify the ambiguity left by Marx regarding the actual meaning of 'accumulation of capital', Luxemburg's work is a significant step forward for an understanding of secular capitalist crisis and is also another indication that Marxism is a living project, both theoretically and in practice.

PENETRATION OF NON-CAPITALIST RELATIONS OF PRODUCTION

The *Communist Manifesto* refers to the 'cheap price of commodities which battered down all Chinese Walls', perhaps leading one to think of an economic process in the narrow sense. However, by the time of *Capital, Volume 1*, Marx stressed, for eighteenth-century England, the coercive, non-economic methods predominating in the formation of wage-labour and how bloody was the history of merchant capital.

Marx refers to the destruction of non-capitalist relations of production in the rise of capitalism from feudalism as 'primitive accumulation'. Yet, such destruction still obtains within the 'regular' accumulation of capital, although the word 'primitive accumulation' would no longer correspond to Marx's usage.[6] One striking and educative illustration is the work of van Onselen (1976) for the formation of the Southern Rhodesian wage-labour force at the turn of the twentieth century. By controlling travel passes, taking advantage of famine, making exaggerated claims about labour conditions in mines, and even using direct force, capitalism drove the rural population of Malawi down a proscribed path: produce your sons for work, or lose your land – that property threatened at the point of the gun if earnings from wage labour were not supplied in sufficient quantity to pay high cash land taxes. Compulsion of this type continues in many manifestations today and is an important dimension of any thorough study of crises of capitalism.

SECULAR 'CRISIS' OF CAPITALISM

Luxemburg once said that if 'capitalist development does not move in the direction of its own ruin, then socialism ceases to be *objectively necessary*' (1899, p. 40, emphasis added). Such an objective necessity for ruin was, for her, one basis for socialism, with two others being the progressive socialisation of production, and increasing working-class organisation and consciousness. The importance of such objective necessity is its addition to subjective factors. It becomes, however, an urgent question after concluding that Marx does not offer a crisis theory.

Luxemburg's *Accumulation* directly deals with crisis only in its very last paragraph. There, she says that capitalism 'strives to become universal, and indeed, on account of this tendency, it must break down – because it is immanently incapable of becoming a universal form of production' (1913, p. 467).[7] In other words, if accumulation of capital represents the extension of capital into non-capitalist portions of a society or the world, it is of decisive importance to know the implications of the very success of such endeavours, even as we understand both the 'economic' and non-economic, forcible methods by which this extension is accomplished. If accumulation of capital is absolutely basic for understanding capital, so too are the implications of its achievements, the implications of capital 'winning' the world in its own name and no longer having that much use for surplus value for additional accumulation of capital.

The Depression of the 1930s can be seen as a consequence of the very success of earlier capitalism as a major crisis of overproduction occurred. The 'solution' was the development of massive amounts of 'unproductive labour' which is wage labour which does not produce value and surplus value (see Chapter 2). Unproductive labour initially developed on a large scale with regard to the implementation of fascism, then as a consequence of the Second World War, followed by Cold War militarism and other developments of unproductive labour. But this is not accumulation of capital. Rather, the development of unproductive labour is a systemic message that the limits of the accumulation of capital are being reached. As such, the fall in the accumulation of capital, relative to the mass of appropriated surplus value, is the deeper message of the past century and fully consistent with Luxemburg's work.

REFERENCES AND FURTHER READING

Bleaney, M.F. (1976) *Underconsumption Theories: A History and Critical Analysis*. New York: International Publishers.

Clarke, S. (1994) *Marx's Theory of Crisis*. New York: St. Martin's Press.

Desai, M. (1991) 'Underconsumption', in T. Bottomore (ed.) *A Dictionary of Marxist Thought* (2nd ed.). Oxford: Blackwell, pp. 552–4.

Hilferding, R. (1910) *Finance Capital: A Study of the Latest Phase of Capitalist Development*. London: Routledge & Kegan Paul, 1981.

Kenway, P. (1987) 'Crises', in J. Eatwell, M. Milgate and P. Newman (eds.) *The New Palgrave: A Dictionary of Economics*. London: Macmillan, reprinted by the same publisher in *Marxian Economics*, 1990, pp. 110–14.

Lapides, K. (1998) *Marx's Wage Theory in Historical Perspective*. Westport, Conn.: Praeger.

Luxemburg, R. (1899), 'Reform or Revolution' (2nd edn., 1908), in *Rosa Luxemburg Speaks*. New York: Pathfinder, 1970, pp. 33–90.

Luxemburg, R. (1913) *The Accumulation of Capital*. New York: Monthly Review Press, 1951.

Marx, K. (1867, 1894) *Capital*, vols. 1 and 3. Moscow: Progress Publishers, 1954, 1959 and London: Lawrence and Wishart, 1974.

Marx, K. (1905, 1910) *Theories of Surplus Value*, parts 2 and 3. Moscow: Progress Publishers, 1968 and 1971.

van Onselen, C. (1976) *Chibaro: African Mine Labour in Southern Rhodesia, 1900–1933*. London: Pluto Press.

Zarembka, P. (2000). 'Accumulation of Capital, Its Definition: A Century after Lenin and Luxemburg', in P. Zarembka (ed.), 'Value, Capitalist Dynamics and Money', *Research in Political Economy* 18. Stamford, Conn. and Amsterdam: JAI/Elsevier, pp. 183–241.

NOTES

1. Clarke (1994, pp. 39 ff.) shows, however, that Hilferding's work on crisis, while quite an advance over bourgeois theories of the time, was based on imperfect competition and without 'specific reference to the social relations of production, so that in the last analysis it was not clear what (if anything) was specifically Marxist in his theory'.

2. Note that, if we want real numbers of the 'rate of profit', we need to remember that surplus value includes all three of profits, interest and rent, and the distribution among these three changes over time.

3. Actually, even Marx (1867, pp. 564–5) has examples for which this is not true.

4. Also, 'the economists ... transform capital from a relationship into a thing, a stock of commodities ... which, insofar as they serve as conditions for new labour, are called capital' (Marx 1910, p. 272, in discussing accumulation of capital).

5. Thus, citing this passage, Clarke (1994, pp. 277–8) notes that the 'ultimate limit of the stimulation of capital accumulation by the expansion of credit is set by the market for final consumption'.

6. 'Accumulation merely presents as a *continuous process* what in *primitive accumulation* appears as a distinct historical process' (Marx 1910, p. 272).

7. Rather than being an underconsumptionist, Luxemburg pointed to the contradiction between expanding production and limited markets for both consumer goods and means of production. Bleaney (1976, ch. 9) correctly understands Luxemburg in this regard, although not grasping her larger point. In his comprehensive book on crisis theory we have otherwise favourably mentioned, Clarke (1994) unfortunately considers Luxemburg to be an underconsumptionist (pp. 53–8), leading him to claim mistakenly that she replaces unlimited expansion of production and productive forces with the notion that 'the development of the market and the growth of consumption ... is the driving force of capitalism' (p. 75), thereby even providing a foundation 'for the reabsorption of Marx's economics back into the framework of bourgeois economic theory' (p. 280). We point out this misinterpretation of Luxemburg in order to alert a reader turning to this aspect of Clarke's work.

16   Marxian Crisis Theory and the Postwar US Economy

Fred Moseley

In the first 30 years after the Second World War, the United States economy performed remarkably well. The rate of growth averaged 4–5 per cent a year, the rate of unemployment was seldom above 5 per cent, inflation was almost non-existent (1–2 per cent a year), and the living standards of workers improved substantially (the average real wage, or the purchasing power of wages, roughly doubled over this period). This was the 'golden age' of US capitalism.

However, this 'golden age' ended in the 1970s. Since then, the rate of growth has averaged 2–3 per cent, the rate of unemployment and the rate of inflation have both been higher, and the average real wage has not increased at all (and by some measures has even declined 10 per cent). It is in this sense that we refer to the 'stagflation' of the US economy in recent decades.

During the late 1990s, the US economy improved significantly, with the highest rates of growth (3–4 per cent) and the lowest rates of unemployment and inflation since the 1960s, and real wages increased modestly. As a result, most economists concluded that the late 1990s 'boom' marked the end of the long period of stagflation and the beginning of a new prolonged period of sustained prosperity, similar to the early postwar 'golden age'. However, this 'boom' came to a sudden end in 2001, and the US economy has fallen again into recession. Now there is widespread concern that this recession will be deep and long, and that it will be accompanied by the first worldwide recession since the 1930s.[1]

This chapter presents a Marxian explanation of the long period of stagflation in the US economy, and attempts to determine whether or not this period of stagflation is indeed over, or whether the US (and world) economy is instead headed for something even worse.

THE DECLINE IN THE RATE OF PROFIT

According to the Marxian theory presented here, the most important cause of the long period of stagflation in the US economy was a very significant decline in the rate of profit (the ratio of total profit to the total capital invested) in the economy as a whole.[2] According to Marxian theory, the rate of profit is the main determinant of the overall condition of capitalist economies. When the rate of profit is high, capitalist economies are generally more prosperous: business investment is high, unemployment is relatively low, and workers' living standards increase (such as occurred in the early postwar 'golden age'). On the other hand, when the rate of profit is low, prosperity turns into stagnation and depression: business investment is low or non-existent, unemployment is high and living standards decline (such as has occurred in recent decades, and occurred during the Great Depression and the nineteenth-century depressions).

From 1950 to the mid-1970s, the rate of profit in the US economy declined almost 50 per cent, from around 22 per cent to around 12 per cent (see Figure 16.1; see Moseley 1991 for a description of the sources and methods used to derive these estimates). This significant decline in the rate of profit appears to have been part of a general worldwide trend during this period, affecting all major capitalist economies.

Figure 16.1   The Rate of Profit in the Postwar U.S. Economy

According to Marxian theory, this very significant decline in the rate of profit was the main cause of both of the 'twin evils' of higher unemployment and higher inflation, and hence also of the lower real wages, of recent decades. As in periods of depression of the past, the decline in the rate of profit reduced the rate of business investment, which in turn has resulted in slower growth and higher rates of unemployment. One important new factor in the postwar period is that many governments in the 1970s responded to the higher unemployment by adopting expansionary Keynesian policies (more government spending, lower taxes, lower interest rates) in attempts to reduce unemployment. However, these government policies to reduce unemployment generally resulted in higher rates of inflation, as capitalist firms responded to the government stimulation of demand by raising their prices at a faster rate in order to restore the rate of profit, rather than by increasing output and employment.

In the 1980s, financial capitalists revolted against these higher rates of inflation, and have generally forced governments to adopt restrictive policies (less spending, higher interest rates). The result was lower inflation, but also higher unemployment. Therefore, government policies have affected the particular combination of unemployment and inflation at a particular time, but the fundamental cause of both of these 'twin evils' has been the decline in the rate of profit.

It is striking that mainstream explanations of the stagflation of recent decades have completely ignored the very significant decline in the rate of profit. These mainstream explanations stress 'exogenous shocks' (i.e. accidents), such as government policy mistakes, the OPEC oil price increase, a mysterious slowdown in productivity growth, etc. According to Marxian theory, all these factors are not 'exogenous shocks', but are instead themselves caused by the decline in the rate of profit. By ignoring the rate of profit, mainstream explanations miss this fundamental cause and remain on the level of superficial appearances.

ATTEMPTS TO INCREASE THE RATE OF PROFIT

Capitalist enterprises have responded to the decline in the rate of profit by attempting to restore it in a variety of ways. We have already mentioned the strategy of inflation, i.e. of increasing prices at a faster rate. Businesses have also attempted to slow down wage increases, and in some cases even to cut wages. Another strategy to reduce wage costs has been to move production operations to low-wage areas of the world. This has been the main driving force behind the so-called 'globalisation' of recent decades: a worldwide search for lower wages in order to increase the rate of profit.

Another strategy has been to make workers work harder and faster. Such a 'speed-up' in the intensity of labour increases the value produced by workers and therefore increases profit and the rate of profit. The higher unemployment of this period contributed to this 'speed-up', as workers have been forced to compete with each other for the fewer jobs available by working harder. One common business strategy has been 'downsizing', e.g. by laying off 10–20 per cent of a firm's employees and then requiring the remaining employees to do the work of the laid-off employees. This method also generally increases the intensity of labour even before the workers are laid off, as all workers work harder so that they will not be among those who are laid off.

We can see that the strategies of capitalist enterprises to increase their rate of profit in recent decades have in general caused suffering for workers – higher unemployment and higher inflation, lower living standards, and increased stress and exhaustion on the job. Marx's 'general law of capitalist accumulation' – that the accumulation of wealth by capitalists is accompanied by the accumulation of misery of workers – has been all too true in recent decades.

However, the startling fact is that, despite the decline in real wages and the 'speed-up' of workers' labour, the rate of profit in the United States has not increased very much since the 1970s (see Figure 16.1). There have been cyclical increases in the rate of profit, especially in the 1990s, but most of these increases have been wiped out in the subsequent downturn, so that overall the rate of profit has recovered only about a third of its previous decline. The rate of profit at the time of writing (2002) remains about 30 per cent below the early postwar peaks. This absence of a full recovery in the rate of profit is the main reason why the US economy has not returned in recent decades to the more prosperous conditions of the 'golden age'. My guess is that the same conclusion also applies to other advanced countries.

Therefore, the most important questions to be answered in a further analysis of the causes of the economic stagflation of recent decades have to do with the rate of profit: What were the causes of the significant decline in the rate of profit in the early postwar period? In recent decades, why hasn't the rate of profit increased more, as a result of the stagnant real wages and the 'speed-up' of labour? And, finally, what is the likely trend in the rate of profit in the future? What are the chances of a significant increase in the rate of profit, which would make possible a full and lasting recovery from the current stagflation and a return to the more prosperous conditions of the early postwar period?

In attempting to answer these key questions about the trends in the rate of profit, mainstream economic theories are of no help, because these theories generally ignore the rate of profit. Mainstream macroeconomics has no theory of profit at all (profit is not a variable in this theory of a capitalist economy!). In microeconomics, the marginal productivity theory of profit (or interest) is completely static (i.e. it provides no theory of trends over time) and is also now in general disrepute, because it has been shown to be logically contradictory (as a result of the 'capital controversy'). This much maligned theory is being quietly dropped from microeconomic textbooks at both the undergraduate and the graduate level.

The only economic theory that provides a substantial theory of the rate of profit and its trends over time is Marxian theory. Indeed, the rate of profit and its trends over time is the *main question* of Marxian theory. The rate of profit is the main variable in Marxian theory, in striking contrast to mainstream theories in which profit is not a variable at all.

Therefore, if we want to understand the causes of the decline in the rate of profit and its likely trend in the future, the only economic theory available to us is Marxian theory. It is often said these days that Marxian theory is 'dead' or 'obsolete'. But this assertion is simply false. As this book itself demonstrates, there are many excellent Marxian economists around the world using Marxian theory to analyse and understand contemporary capitalism, including the current world economic crisis. Indeed, Marxian theory is essential if we want to understand the rate of profit and its trends. There is simply no credible alternative theory of the rate of profit available.

Let us turn now to the explanation offered by Marxian theory of the decline in the rate of profit in the postwar US economy, and of the lack of a full recovery of the rate of profit in recent decades.

MARXIAN THEORY OF THE DECLINE IN THE RATE OF PROFIT

The main point of the Marxian theory of profit is that profit is produced by workers, by the surplus labour of workers, because the value added to commodities by the labour of workers is greater than the wages the workers are paid (profit is equal to the difference between the value produced by workers and the wages they are paid). This conclusion follows from the labour theory of value, which is usually misinterpreted by mainstream economists as a theory of individual prices, like mainstream microeconomics. But this is a misunderstanding. The Marxian labour theory of value is mainly a macroeconomic theory of the total profit produced in the economy as a whole.

Marxian theory concludes that the rate of profit (the ratio of the total profit to the total capital invested) will decline over time, because technological change – an inherent, ever-present feature of capitalist economies – tends to replace workers with machines, and thus tends to reduce the number of workers employed in relation to the total capital invested in machinery, etc. However, since profit is produced by workers, the reduction in the number of workers employed also reduces the amount of profit produced, in relation to the total capital invested. In other words, the rate of profit will decline. Expressed inversely, technological change causes the total capital invested to increase faster than the number of workers employed, or causes the average *capital invested per worker* to increase, which in turn causes the rate of profit to fall.

Marxian theory argues further that the negative effect on the rate of profit of the increase in the capital per worker can be partially offset by increasing the amount of *profit produced by each worker*, which also tends to increase as a result of technological change, which increases the productivity of labour. This positive effect of new technology and higher productivity on the profit produced per worker is also reinforced by other ways of increasing the profit per worker, such as wages cuts and increases in the intensity of labour, discussed above.

However, Marxian theory argues that there are inherent limits to the increase in the profit produced by each worker. The main limit is that there are only so many hours in the working day, and it becomes harder and harder to increase the profit produced by each worker in a given working day. Another limit is the resistance of workers, who usually fight against wage cuts and fight for higher wages and a share of the benefits of the higher productivity. As a result of these limits, Marxian theory concludes that 'labour-saving' technological change will eventually cause the rate of profit to decline. This decline in the rate of profit is no accident, nor is it due to 'external causes'. Rather, the decline in the rate of profit is the result of capitalism's own internal dynamics characterised by continual technological change (see Moseley 1991, ch. 1, for a further discussion of Marx's theory of the falling rate of profit).

The above trends, predicted by Marxian theory, are essentially what happened in the postwar US economy. Technological change increased the capital invested per worker, and also increased the amount of profit produced by each worker. And, as predicted by Marxian theory, the capital invested per worker increased faster than the profit produced per worker, so that the rate of profit declined significantly, as we have seen above.

Another important determinant of the rate of profit, according to Marxian theory, which Marx himself did not emphasise, but which seems to have been important in the postwar US economy, is the ratio between *productive labour and unproductive labour* in the capitalist economy. According to Marxian theory, profit is not produced by all employees in capitalist firms, but only by workers engaged directly or indirectly in production activities (actually making or designing or transporting something), which Marx called 'productive labour'. There are two other main groups of employees who are *not* engaged in production activities, which Marx called 'unproductive labour': 'sales' employees (sales and purchasing, accounting, advertising, finance, etc.) and 'supervisory' employees (managers, supervisors, 'bosses' in general). These two groups of unproductive labour, although entirely necessary within capitalist firms, nonetheless do not themselves produce value and profit (see Chapter 2 and Moseley 1991, ch. 2, for a further discussion of Marx's concepts of productive and unproductive labour).

According to Marxian theory, if unproductive labour (which does not produce profit) increases faster than productive labour (which does produce profit), this will also cause the rate of profit to fall, because costs are increasing, but profit is not, for the economy as a whole. This is what happened in the postwar US economy: the ratio of unproductive labour to productive labour almost doubled during the 'golden age', and this very significant increase contributed to the decline in the rate of profit. This increase in the ratio of unproductive labour to productive labour also seems to have been due in large part to technological change, which increased the productivity of production workers more rapidly than that of non-production workers, and which therefore required more and more sales workers to sell the more rapidly increasing output of production workers (see

Moseley 1991, ch. 5, for a further discussion of the causes of the relative increase of unproductive labour).[3]

Therefore, according to the Marxian theory presented here, there were two main causes of the decline of the rate of profit in the postwar US economy from the late 1940s to the mid-1970s: an increase in the capital invested per worker, and an increase in the ratio of unproductive labour to productive labour. According to my estimates, these two trends contributed roughly equally to the total decline in the rate of profit during this period (see Moseley 1991, ch. 4). Both of these causes were themselves the result of technological change, an inherent feature of capitalist economies. Therefore, the decline of the rate of profit in the postwar US economy was not due to accidental, external causes ('exogenous shocks'), but was instead due to the inherent dynamic of technological change. It is an interesting and important question whether this Marxian explanation of the stagflation of recent decades also applies to other advanced countries. My conjecture is that it does.[4]

WHAT IS NECESSARY TO INCREASE THE RATE OF PROFIT?

What does the Marxian theory presented above imply about what must be done in order to increase the rate of profit, and thereby make possible a revival of capital investment and a return to the more prosperous conditions of the 'golden age'? According to this Marxian theory, the rate of profit varies directly with the profit per worker, and varies inversely with the capital per worker and the ratio of unproductive labour to productive labour. Therefore, there are three main ways to increase the rate of profit: (a) increase the profit per worker, (b) reduce the capital per worker, and (c) reduce the ratio of unproductive labour to productive labour.

Marxian theory suggests further that an increase in the profit produced per worker (by means of wage-cuts, speed-ups, etc.) is not likely by itself to be sufficient to restore the rate of profit to its previous levels, since the prior decline in the rate of profit was not caused by a decline in the profit per worker, but was instead caused by increases in the capital per worker and in the ratio of unproductive labour to productive labour. We have already seen that a significant increase in profit per worker in recent decades has resulted in a relatively small increase in the rate of profit. Marxian theory suggests that what is required to fully restore the rate of profit is to reverse the two trends that caused its decline, i.e. to reduce the capital invested per worker and to reduce the ratio of unproductive labour to productive labour.[5]

The main way capital per worker has been reduced in the past has been through the widespread bankruptcies of capitalist firms, which are caused by the combination of falling profits and rising debts. As a result of bankruptcies, surviving firms are able to purchase the productive assets of the bankrupt firms at a very low price, thereby reducing the amount of capital invested per worker and raising their rate of profit. This process of bankruptcies, etc. (which Marx called the 'devaluation of capital') continues until the capital per worker has been reduced enough and the rate of profit increased enough in the economy as a whole for capital investment to resume and for a period of recovery and expansion to begin. Of course, widespread bankruptcies also worsen the economy in the short run, and many times in the past have turned a recession into a depression.

The main way to reduce the ratio of unproductive labour to productive labour would be to lay off large numbers of non-production employees (sales, managers, etc.). Leaving aside the questions of whether such a large reduction of non-production employees is feasible in the US economy today, and how it would be brought about, such a large displacement of non-production employees would sharply increase the rate of unemployment, especially among these occupations. Thus we can see that all the various ways in which the rate of profit could be increased (wage-cuts, bankruptcies, lay-offs, etc.) involve hardships and declining living standards for workers.

Since the mid-1970s, as discussed above, profit per worker has increased significantly (through wage cuts, etc.) and this has indeed contributed to an increase in the rate of profit (while at the same time contributing to an increase in the hardships of workers). However, the other two crucial adjustments necessary to increase the rate of profit have not yet happened in the US economy. The capital invested per worker has remained more or less constant (first decreasing in the 1980s and then increasing in the 1990s) and the ratio of unproductive labour to productive labour has continued to increase (although at a slower rate) and thus has continued to have a negative effect on the rate of profit. This is the main reason why the rate of profit has increased so little since the 1970s, in spite of the significant increase in the profit produced per worker (see Moseley 1997 for a further discussion of the trends in these key variables since the 1970s).

WHAT LIES AHEAD?

What does this Marxian theory imply about the future course of events in the US (and world) economy? In the first place, this theory implies that the future of the US economy, like its past, will depend mainly on the rate of profit. If the rate of profit increases significantly, then perhaps the US economy will return to the more prosperous days of the early postwar 'golden age'. However, if the rate of profit remains at the current low levels, then a return to prosperity is not very likely. Instead, the US economy will continue to experience sub-par growth and higher unemployment, and perhaps even worse.

Furthermore, this theory suggests that the future trend of the rate of profit depends on the three main factors discussed above: the profit produced per worker, the capital invested per worker, and the ratio of unproductive labour to productive labour. Profit per worker will probably continue to increase as in recent years (as slow growth and higher unemployment continue to put downward pressure on wages), which will continue to have a positive effect on the rate of profit. Further, if the economy continues to expand (although slowly), then the capital per worker will probably also increase slowly, which will have a negative effect on the rate of profit. And the ratio of unproductive labour to productive labour will probably continue to increase, which would continue to have a negative effect on the rate of profit. The net effect of these opposing trends is difficult to predict with precision, but extrapolating from the recent past, it does not appear very likely that there will be a significant increase in the rate of profit in the foreseeable future. In the absence of such an increase in the rate of profit, the US economy will at best remain stuck in the stagnation of recent decades (see Moseley 1999 for a further discussion of the likely future trends in these key variables).

Furthermore, according to Marxian theory, there is not much the government can do to avoid this gloomy prospect, because there is not much that government economic policies can do to increase the rate of profit. Expansionary fiscal and monetary policies do not increase the profit produced per worker, nor reduce the capital invested per worker, nor reduce the relative proportion of unproductive labour. Therefore, even though expansionary and monetary policies may provide a temporary boost for the economy, they are not able to achieve the necessary precondition for a return to a new era of prosperity: a significant increase in the rate of profit.

As this is written, six months after 11 September 2001, the US economy is falling again into recession, which threatens to be severe. This recession has not been caused by the September attacks, but rather by a rapid decline in the rate of profit since 1997 (see Figure 16.1), which has led to a sharp reduction in capital investment beginning in 2000, and then to a spreading recession in recent months (even before 11 September). Also, capitalist firms contracted record amounts of debt in recent years in order to finance the 'investment boom' of the late 1990s (and also to help finance the stock market boom of the late 1990s, as firms used about half of the money they borrowed to repurchase their own stock!). This combination of low profits and high debt makes the risk of defaults and bankruptcies today the highest in the postwar period. Also, households have taken on record levels of debt to finance their 'consumer spending spree' of the late nineties, and they too face a higher danger of defaults and bankruptcies, which would further worsen the recession.

Another important aspect of the current recession is that the other two major economies in the world are also either in recession (Japan, which has been mostly in recession for the last decade), or falling into one (Europe). This synchronised recession of the three major economies will in turn have devastating effects on the rest of the world, especially developing countries, who depend very much on exports to the three main economies. Most of the rest of the world – Asia, Latin America and Africa – is already in a deepening recession, as a result of the slowdown of the major economies (e.g. Mexico, which sells 85 per cent of its exports to the United States, has been in a recession for over a year). The 'Asian crisis' has returned, this time without the capital flight and currency crises, but probably even worse in terms of GDP declines and higher unemployment, because the major economies are also in a recession this time (there is no 'US locomotive' to pull these countries out of recession this time). This will be the first global recession in the world economy as a whole since the Great Depression.

Whether the current global recession turns into global depression cannot be predicted with precision. But if Marx's theory (and history) is any guide, the postwar period of declining profitability and increasing debt will eventually be followed by a period of depression, characterised by significant and widespread bankruptcies which will eventually raise the rate of profit for surviving firms and eliminate much of the existing debt, thereby creating the conditions for another period of expansion and prosperity. In other words, a return to prosperity requires a prior depression. It may be possible to continue to avoid such a depression for a few more years; but without such a depression, Marxian theory suggests that a return to the more prosperous conditions of the early postwar 'golden age' is not very likely.

Such a worsening crisis of global capitalism would inflict great suffering – loss of jobs, lower incomes, greater hunger and poverty, greater anxiety and desperation, etc. – on the world's working population, especially in developing countries. How would workers around the world and in the United States respond to this widespread and increasing misery? In seems likely that in the next few years workers all over the world will be forced to choose between passively accepting higher unemployment and lower living standards or actively resisting these hardships and striving to defend their economic livelihood. It is possible that, as economic conditions deteriorate, these struggles by workers to maintain their living standards within a capitalism in crisis will lead more and more of them to call into question capitalism itself, and the adequacy of capitalism to meet their basic economic needs. If capitalism requires these attacks on our economic livelihood, then perhaps there is a better economic system that does not require such attacks and which could better satisfy our economic needs and wants.

REFERENCES AND FURTHER READING

Freeman, A. (1991) 'National Accounts in Value Terms: The Social Wage and the Profit Rate in Britain, 1950–1986', in P. Dunne (ed.) *Quantitative Marxism*. Cambridge: Polity Press.

Moseley, F. (1991) *The Falling Rate of Profit in the Postwar United States Economy*. London: Macmillan.

Moseley, F. (1997) 'The Rate of Profit and the Future of Capitalism', *Review of Radical Political Economics* 29(4), pp. 23–41.

Moseley, F. (1999) 'The US Economy at the Turn of the Century: Entering a New Era of Prosperity?', *Capital and Class* 67, pp. 25–45.

NOTES

1. The rest of the world has experienced a similar deterioration of economic performance in recent decades. The rate of unemployment has been above 10 per cent in Europe for most of the 1990s. Japan has been in a prolonged economic slump for over a decade. The other Asian economies have also fallen into crisis in recent years. Economic conditions have been especially severe in Latin America and Africa, which have suffered two

'lost decades', in which growth has been stagnant and living standards have declined drastically.

2. Total profit includes the interest paid to creditors, and hence is a comprehensive measure of the total 'return to capital' for capital as a whole, including both non-financial and financial capital.

3. In Chapter 15, Paul Zarembka presents a very different interpretation of the effects of an increase of unproductive labour. Zarembka argues that an increase of unproductive labour *solves the problem of the realisation of surplus value* which is inherent in capital accumulation. In my interpretation, for which there is considerable textual evidence (see Moseley 1991, ch. 2), unproductive labour is a cost, a *deduction from surplus value*. Therefore, an increase of unproductive labour, in relation to productive labour that produces surplus value, increases the relative deductions from surplus value, and hence causes the net rate of profit going to capitalists to decline. In other words, a relative increase of unproductive labour is an important *cause* of capitalist crises, not a *solution* to capitalist crises. Zarembka also argues that the question of a secular decline in the rate of profit is no longer of theoretical interest: 'And, while falling profits in the first half of the nineteenth century encouraged economists to see this as a fact requiring theoretical explanation, a century and a half later of rising, falling, rising, falling (etc.) rates of profit pushes into the background even searching theoretically for a tendency in profit rates.' However, the all-important fact is that the rate of profit in the US economy (and other major capitalist countries) *declined significantly* (about 50 per cent) in the 1960s and 1970s, and has not recovered since. I have argued above that this significant decline in the rate of profit was the main cause of the deterioration of economic performance since the 1960s. Therefore, the explanation of this significant decline in the rate of profit is a very important question, both theoretically and practically. It may be that Marx's theory cannot prove that there is a long-term, secular decline in the rate of profit over centuries (I don't think that is what Marx was trying to do). But Marx's theory does provide an explanation of the long-term decline in the rate of profit in the postwar US economy (and I think other economies as well). According to this Marxian explanation, the decline in the rate of profit was caused in part by the relative increase of unproductive labour. Therefore, the relative increase of unproductive labour in the postwar US economy was a big *problem* for capitalism, not a *solution* to the realisation problem.

4. For a similar explanation of the decline in the rate of profit in the postwar UK economy, see Freeman (1991).

5. Japan has been in a prolonged slump for the entire decade of the 1990s, and has fallen again during 2002 into even deeper recession. According to Marxian theory, the main reasons for this prolonged slump are: a significant decline in the rate of profit and the unwillingness (or inability) of Japanese banks to force money-losing firms into bankruptcy. The Japanese experience seems to suggest, in support of Marxian theory, that these bankruptcies cannot be avoided.

17 Where is Class Struggle?

John Holloway

WHERE IS CLASS STRUGGLE TODAY?

Perhaps the commonest argument against Marxist theory today is not that it is wrong in its criticism of capitalism but that it is wrong in insisting on the importance of class struggle. The evils of capitalism are plain for all to see, but where is the class struggle that Marxists keep talking about? Struggle certainly, struggle there is: the struggles of the anti-globalisation movement from Seattle to Genoa, the struggles of the Zapatistas in Mexico or of the landless peasants in Brazil, the struggles of women and gays against discrimination, the struggles to protect the environment, even the spectacular protest of the people who flew the planes into the World Trade Center. Struggle is easy to see, but is it class struggle? Where has the labour movement been in the last 20 or 30 years? Certainly not leading the revolution. Is it not better then to stop talking about class struggle and to speak of new social actors or simply of 'the multitude'?

This chapter argues that class struggle has probably never been so vicious and violent as it is today.

THE EXISTENCE OF CAPITAL IS CLASS STRUGGLE

But no, we must go back before that. Why start with capital, why not with racism or patriarchy? Are there not lots of different types of oppression?

Let's not start with capital, then. Let's start with ourselves. We want to change the world. We want to change the world because it stinks, because it is obscenely unjust, because it is violent, because, the way society is organised at the moment, it looks as if humanity will probably destroy itself before very long. (If you don't want to change the world, go and read a different book.)

Changing the world implies doing. If we want to change society, then we must think of society not as-it-is, but as something that people have made and that people can change. We must think of ourselves as doers, not as beings. This is sometimes referred to as materialism, or even dialectical materialism: what Marx means by materialism is basically understanding society in terms of human doing, 'sensuous human activity, practice' (Marx 1976, p. 3, 'First Thesis on Feuerbach').

Doing is central to any revolutionary project, to any project of changing the world. That is why revolutionary thought means thinking of society in terms of doing. This is not just a call to action, to rushing out in the street and doing something. It means first trying to understand society (and our own repudiation of present society) in terms of the way that human doing, human activity, is organised.

When we think of our doing, our activity, one of the first things that strikes us is that it is social. It is difficult to think of any activity that does not depend on the doing of others, either now or in the past. I sit here writing and think of it perhaps as my great individual act, but I know that without the doing of the people who made the computer, installed the electricity, built the building, made the desk, wrote all those other books that have influenced me, I could not be writing what I am writing. The doing of others is always the precondition of our own doing, the means of our doing; and our doing becomes the means of the doing of others. Our doing is always part of a social flow of doing.

Our doing is not only social, it is also projective. An aspect of our doing is the aim to change, to make things other than they are. Our doing includes a projection beyond that which is. This projection-beyond is fundamental to any idea of changing the world. For Marx, it was also the characteristic that distinguished humans from animals – or from machines, we might add. In a famous passage in *Capital*, he contrasts the architect and the bee: 'A spider conducts operations that resemble those of a weaver, and a bee puts to shame many an architect in the construction of her cells. But what distinguishes the worst architect from the best of bees is this, that the architect raises his structure in imagination before he erects it in reality. At the end of every labour-process, we get a result that already existed in the imagination of the labourer at its commencement. He not only effects a change of form in the material on which he works, but he also realises a purpose of his own' (1965, p. 178).

What happens to doing in capitalism? Somebody comes along, takes what we have done and says 'this is mine'. The capitalist (and the feudal lord and the slave-owner before him, but let us focus on the capitalist) appropriates that which we have done, the product of our work (see Chapter 1).

This shatters doing. The capitalist, by appropriating the product, shatters the social flow of doing. (Oh, he may conceivably be a very nice man, or possibly woman, and this is not a very nice thing to do, but if he doesn't do it, he won't be a capitalist: that's what Marx means by talking of capitalists as the 'personifications of capital'.) What we have done becomes his property. He now owns the means of doing of others – the means of production, in other words. In order for others to do (and to survive) they must get access to the means of production: this they do by selling their labour power (their capacity to do) to the owner of the means of production. Their doing is now converted into labour at the command of the capitalist, labour which produces commodities which are, of course, the capitalist's property and which he can sell on the market.

The direct connection between our doing and the doing of others is broken. Of course, our doing is still part of a social flow of doing, but it does not appear that way. Now my writing appears as an individual act in which I use a number of *things* (computer, electricity, desk, and so on), for which I have paid money, commodities which I buy on the market. Our doing is social, but is not directly social: it is indirectly social.

The social flow of doing is broken, and so too is the projective character of doing. It is now the capitalist, the owner of the means of production to whom the worker sells her labour power, who decides what the worker will do. The conception or projection is now the activity of the capitalist, the execution is left to the worker. With that, the worker is reduced to the status of a bee, having been deprived of the projection which is the distinctive character of humans. This does not just happen once, it is a repeated process. Since the capitalist appropriates the product of the labour, the worker leaves the labour process just as poor as she entered, just as dependent on the kindness of the capitalist in employing her the next day. The capitalist, on the other hand, gets richer and richer, since the product produced by the worker is worth more than what the capitalist has paid her for her labour power. The capitalist, in other words, exploits the worker, extracts more value from the worker than he pays her.

SO WHAT?

We started with interesting things like the Zapatistas and the World Trade Centre, and now the argument seems to be dragging us into the factory, dragging us towards some sort of economic theory of society. That's the problem with Marxism, isn't it?

No, no, you didn't understand. Or perhaps I didn't explain. Doing is not in itself economic. It is lying in bed, making dinner, eating, writing articles, doing essays, making love, going on a demonstration, making a chair, producing electricity, making a car, whatever. It is the existence of capitalism that defines a major part of our doing as labour, as an 'economic' activity. It is the fact that our only access to the means of survival is through money and our only way of getting money is by selling our labour power and turning a big part of our daily doing into labour at the command of others – that is what puts 'economics' at the centre of our lives. But though the critique of capitalism is not an economic critique, it is necessarily a critique *of* economics. That is why Marx in *Capital* did not develop an alternative economics but a critique of economics (or a critique of political economy). The struggle is to emancipate doing from labour, life from economics. That is why our analysis cannot start from labour, but must begin with doing, 'sensuous human activity, practice'.

Doing, then, in capitalism, is a shattered, fragmented, separated process. 'Separation', says Marx, is the 'real generation process of capital' (1972, p. 422). Capital is the separation of the vast majority of people from the means of doing (means of production), the separation of the product from the producers, the separation of people from purposive social activity (their 'species-being', as Marx puts it in the *1844 Manuscripts*), the separation of people from one another, the separation of 'labour' from other forms of activity. This separation affects absolutely every aspect of our lives. The doing which is separated off from labour is affected, for it is defined as leisure, as secondary, as not very serious – 'I don't do anything, I'm just a housewife', or 'I don't do anything at the moment, I'm unemployed.' The doing which creates the conditions for the doing of others is now seen as producing 'things' for others to buy. The relations between people thus become refracted through things; they become reified or fetishised, as Marx puts it in the first chapter of *Capital*. The way in which we relate to the doing of others is through the exchange of things. Social relations acquire the form of things (such as money, commodities, capital, the state) which we do not recognise as forms of social relations, nor as the product of our own doing. Fetishism does not refer just to the economic, to the appearance of social relations in economic forms, but to every aspect of society, every aspect of the way in which we see our own lives and our relations to others. Every aspect of our doing is transformed by that fundamental rupture of the social flow of doing, that separation process which is capital.

And how does the separation take place? Through struggle, of course. It does not take place automatically. Think of what the capitalist says to us each day. He says 'all of these wonderful things that you see around you, all these things that you would like to have in order to survive, all these things that you would like to enjoy, all of that is private property, all that is mine. If you want to enjoy any of these things, you can do so by earning money. But in order to get money, you must give up any idea of spending your days doing what you like: you must come and labour for me and, if you do what I tell you, then I shall give you money so that you can buy some of these things. But, mind you, I shall only give you enough money to keep you going for a short time, tomorrow you must come back and labour for me again. And of course I shall employ you only as long as I manage to exploit you successfully and make a profit from employing you.' He says all this as though it were obvious, but of course it is not. Think of the poor capitalist. In order to carry through his existence as a capitalist, he must persuade or force us to accept what he says when he says 'this is mine': he must force us to respect his private property. This is not done easily: it requires the services of millions of police and security guards, not to mention teachers, social workers and parents. Then he must persuade or force us to accept the horrifying, absurd idea that we should turn our daily doing into labour under his command. In general terms, he does this by protecting his private property, but he still has to get us up early in the morning (never an easy task, especially when the prospect is going to work at the command of others), and get us to go to work and to do what he commands and do it efficiently (more efficiently than his fellow capitalists, with whom he has to compete).

All this is not easy. It is made more complicated by the fact that we are not slaves or serfs. We are free. 'Ha ha!', you laugh, 'free to obey the command of the capitalist'. Yes, certainly, but also, as a result of the struggles of slaves and serfs, really free in a way that is sometimes inconvenient for the capitalist. He cannot stop us from going and selling our labour power to a different capitalist if we get tired of him. He also cannot (usually) shoot us or flog us if we do not obey his commands at work. In other words, unlike the feudal lords, the capitalists require the support of an apparently external instance in order to apply the sort of violence that is required to maintain a system of exploitation. This is the state. The separation of state and society, the political and the economic, is another, very important aspect of the separation process which is capital. The state, as part of the process of separation, regulates the process of separation as best it can: it protects the separation of done from doing, the 'this is mine!' of private property which ruptures everything. All systems of exploitation are armed robbery; what is peculiar about capitalism is that the person who holds the arms is distinct from the person who does the robbing (see Chapter 9).

All of this is struggle. Of course it is. It is struggle over what we think, how we act, struggle over how we get up in the morning and therefore over what time we go to bed and what we do in bed, it is the struggle of the alarm clock, as we throw it at the wall and then still get up to go and sell our labour power. It is an extremely violent struggle in which thousands and thousands of people die each day, because they are cut off from the flow of doing and starve, because of the repression against those who do not accept the 'this is mine!' proclaimed by the capitalists, because what they are commanded to do by capital is dangerous. It is a struggle that comes in the first place from them, from the capitalists. If it were up to us, we would lie in bed or potter about all day or dedicate ourselves with passion to whatever we like to do. It is they who do not leave us in peace, it is they who say 'get up out of bed and come to labour for us – or starve, if you prefer'. The struggle, then, is between two ways of doing, two forms of social relations. Capital is the imposition on our lives of a certain form of doing, a certain form of relating to one another. Capital is the struggle both to transform our doing into labour and also, once we are in the place of work, to get us to work as capital commands.

And of course we struggle back. We struggle back because we are not yet machines. We struggle back by throwing the alarm clock at the wall, by going to bed late even when we know it will impair our efficiency at work the next day, we struggle back by giving higher priority to playing with our children than to producing profit, we struggle back by fighting for higher wages at work or by fighting for more acceptable conditions, we struggle back when we demonstrate against the misery created by the imposition of private property, we struggle back by projecting beyond capitalism, by dreaming of a better society, a society in which we ourselves decide what we do. We struggle in the workplace and outside the workplace. We struggle for a different way of doing, a different form of social relations.

THIS, THEN, IS CLASS STRUGGLE

It is class struggle not because we wear cloth caps. It is class struggle not because we think of ourselves as being left-wing, but just because we live and want to live: if class struggle were exclusive to the left or the consciously militant, there would be no hope. It is class struggle not because we want to be the working class, but because we don't want to be working class. On our part, it is struggle not for being a class but against being a class. It is capital that classifies us. It is capital that says to us each day, 'you are without property, you must come and work for us; and then you will go home without property and come back the next day, and so on for the rest of your life, and so on for the lives of your children and of their children'. It is capital that ensures that each day what is produced in the factory is not just commodities but two classes. As Marx puts it, 'capitalist production, therefore, under its aspect of a continuous connected process, of a process of reproduction, produces not only commodities, not only surplus value, but it also produces and reproduces the capitalist relation; on the one side the capitalist, on the other the wage-labourer' (1965, p. 578). Capitalist production produces classes, imposes discipline and regimentation, forces our doing into the narrow band of labour that produces profit for the capitalist (or labour that supports capitalist profitability, in the case of state employment). Capital is grey, we are the rainbow, fighting for a world in which doing becomes free, liberated from the bonds of value production, from the chains of profit.

The existence of capital, then, is class struggle: the daily repeated separation of people from the social flow of doing, the daily repeated enforcement of private property, the daily repeated transformation of doing into labour at the command of capital, the daily repeated seizure of the products of that labour. It is class struggle but it does not appear as such. It does not appear to be class struggle (or indeed struggle at all) because of the very nature of the struggle itself. Capital's struggle is to separate doers from the social flow of doing, but that very separation means that people no longer understand themselves as doers, nor as social. The separation means that people appear as individuals and as beings rather than as doers. The more successful capitalist class struggle is, the more invisible it becomes: people are transformed from doers bound together by the community of their doing into free and equal individuals tied together by external institutions such as the state. Capitalist class struggle takes place through apparently neutral forms such as property, money, law, state. These are all forms through which capital as a form of doing is imposed upon our lives. Capital does not say 'we are going to exploit you, we are going to force you to labour for us until you crawl home exhausted at the end of the day, and then we are going to force you to come back and back for the rest of your life'. No, capital simply says a number of key words: 'Respect private property, the money, the law.' When it says 'respect private property', what it means is 'stand aside while we separate you from the means of doing and of survival'. 'Respect money' means 'just let us get on with shattering all social relations, with mediating all relations between people through money'. And respecting the law means that we should give up all idea of shaping our own doing, that all activity must be made to conform with the acceptance of private property, that we must bow to an external force. Capitalist class struggle comes clothed in liberal theory.

Capital cannot stand still. Under feudalism, the relations of exploitation were more or less stable: what the lord demanded of his serfs did not vary very much over time. It is different with capital. Capital is driven forward constantly by its 'were-wolf's hunger for surplus-labour' (Marx 1965, p. 243). The fact that capital is fragmented into many distinct capitalist units (companies), each in competition with the others, each depending for its survival on being able to exploit its workers more effectively than the others, means that capital can never stand still, that it is constantly driven forward to intensify the exploitation of labour. Intensifying exploitation means not only imposing tighter discipline in the factory but creating the conditions in society (that is, the world) that make this exploitation easier. This means subordinating every aspect of life more and more tightly to the aims of value production. More and more intensely, every aspect of life becomes a battleground for the imposition of capital, a battleground on which we resist and try to defend and develop what we consider to be human or emancipatory. In education, for example, recent years have seen a huge assault in all parts of the world to bring teaching into line with the requirements of the market (that is, with capital). Sometimes, this takes the form of outright privatisation, allowing the content of education to be dictated directly by the market, sometimes it means the introduction of various forms of control within the state system of education to achieve the same end. Of course that assault meets constant resistance, either in the form of student strikes or, far less dramatically, in the form of teachers and students pushing beyond market requirements in the attempt to develop an honest and critical understanding. Sometimes it takes the form of teachers writing or recommending and students reading books like this one and suddenly finding that, just when they thought that class struggle was dead, they are in the middle of it.

CAPITAL HAS BEEN PARTICULARLY VIOLENT IN RECENT YEARS

The violence stems from the curse that strikes all who try to dominate and exploit: they depend absolutely upon those whom they exploit and dominate. Capital depends for its existence on its capacity to transform doing into labour and to exploit that labour. Not just that: it depends for its existence, as we have just seen, not just on being able to maintain exploitation but on being able to intensify it all the time. It needs not just to subordinate society, but to subordinate it more and more and more. If it cannot do so, it falls into crisis. The crisis tells capital that in order to survive, it must intensify subordination. In the 1970s, capital was manifestly in crisis, socially and economically. The weakness of subordination was obvious in society from the late 1960s: strikes, demonstrations, student movements, revolutionary movements in many parts of the world. The failure to subordinate became economically manifest in the worldwide crisis which broke in 1973–74. The response of capital came in various forms (emphasis on a return to traditional family values, increasing police and in some cases military action), but above all it took the form of money and property.

Money has been central to class struggle over the last 25 years or so. First, in the early years of Reagan and Thatcher, tight monetary policies were applied as a means of reducing debt and imposing the discipline of the market. When this strategy threatened to destroy the world financial system, a more flexible approach to debt was adopted, allowing more spending to those (basically big companies, rich people and rich states) whose well-being was considered essential to the health of capitalism, while using debt as a means of disciplining those who required discipline (basically the poor) or were simply disposable (a large part of the world's population is not an 'industrial reserve army' for capital in any sense, but simply a nuisance).

The extension of property has also been important to capital's struggle. Just as, in the early days of capitalism, landowners pushed the peasants off the land, enclosed the land and said 'this is mine', so capital is now enclosing more and more areas of human activity and saying 'this is my property, this is mine'. The development of the concept of 'intellectual property' has been crucial in this. Capital spends enormous sums of money on trying to assert its property rights over music, software, pharmaceutical discoveries, genes and so on. In many parts of the world this extension of property rights has been carried out with a remarkable violence, as the traditional agricultural or medicinal knowledge of communities has been appropriated by capitalist enterprises, which patent the knowledge without any compensation and then enforce their property rights against all comers, including the communities themselves.

Money and property are very violent forms of struggle. They have undoubtedly caused far more deaths in the last twenty years than all the wars fought in the last hundred. Each day 35,000 children die simply because property and money separate them from what they need to survive. But they are also remarkably vulnerable forms of struggle. Money, especially in the form of debt, is contested all the time, both by those who will not pay and by those who point to the enormous destruction of human lives that the enforcement of debt involves. Property, especially intellectual property, is also contested almost universally – both by those who habitually copy software, videos and CDs, and by those who campaign against the misery caused to AIDS sufferers, for example, by the protection of pharmaceutical patents. Both of these issues have been important features of the anti-capitalist movement of recent years.

Is this class struggle? Of course it is. Why not refer to it just as a multiplicity of struggles by new social actors or by the 'multitude'? Because the concept of class points to the fact that behind the particular issues (AIDS in Africa, copying music from the internet, student loans, poverty in Latin America) there is a single struggle: the struggle by capital for profit, that is, the struggle by capital to exploit, to convert doing into labour and impose its form of social relations, and the struggle by us against all that, for a different form of doing, a society based on the recognition of human dignity. Class points to that underlying unity in a way that the other categories do not. It also points to something else that is fundamental: there is no certainty that we shall win, that humanity will survive the attacks of capital, but the concept of class gives us hope, for it shows that capital depends upon us for its existence, that we are the only subjects.

REFERENCES AND FURTHER READING

Bloch, E. (1993) *The Principle of Hope*, 3 vols. Oxford: Basil Blackwell.

Bonefeld, W. and Holloway, J. (eds.) (1995) *Global Capital, National State and the Politics of Money*. London: Macmillan.

Dinerstein, A. and Neary, M. (eds.) (2002) *The Labour Debate*. London: Ashgate.

Holloway, J. (2002) *Change the World Without Taking Power*. London: Pluto Press.

Marx, K. (1965) *Capital*, vol. 1. Moscow: Progress Publishers.

Marx, K. (1972) *Theories of Surplus Value*, part 3. London: Lawrence and Wishart.

Marx, K. (1976) Theses on Feuerbach, *Marx and Engels Collected Works*, vol. 5. London: Lawrence and Wishart.

Marx, K. *Capital* (any edition).

Marx, K. *Economic and Philosophic Manuscripts of 1844* (any edition).

Smith, C. (1996) *Marx at the Millennium*. London: Pluto Press.

18 Transcending Capitalism: The Adequacy of Marx's Recipe

Michael Lebowitz

THE VISION OF AN ALTERNATIVE

In the 'Postface to the Second Edition' of Volume 1 of *Capital*, Marx mocked a French reviewer of the first edition who criticised him, he said, for not 'writing recipes ... for the cook-shops of the future' (Marx 1977, p. 99).[1] Although he did not respond here, Marx's answer would be apparent from his comments elsewhere about utopian socialists who merely constructed 'fantastic pictures and plans of a new society'. While not rejecting the goals of the utopians, Marx stressed that the means of achieving them were not through propaganda and exhortation; rather, workers through their struggles would bring about a new society: 'the real conditions of the movement are no longer clouded in utopian fables' (Marx and Engels 1971, p. 166).

Nevertheless, Marx did have a clear conception of an alternative to capitalism. His goal, like that of other early socialists, was the creation of a society that would allow for the full development of human potential and capacity. As his associate Friedrich Engels wrote in an early draft of the *Communist Manifesto* ('Draft of a Communist Confession of Faith'), the goal of the communists is 'to organise society in such a way that every member of it can develop and use all his capabilities and powers in complete freedom and without thereby infringing the basic conditions of this society'. Marx's final version of the *Manifesto* stresses the indivisibility of this goal, calling for 'an association, in which the free development of each is the condition for the free development of all'.

At the very heart of Marx's conception of the society of free and associated producers was the removal of all fetters to human beings – just as the stunting of human potential and the tendency to reduce human beings to beasts of burden and things was at the core of his rejection of capitalism. From his earliest writings, Marx stressed the potential for the development of rich human beings with rich human needs, the potential for producing human beings as rich as possible in needs and capabilities. What, indeed, is wealth, he asked, 'other than the universality of individual needs, capacities, pleasures, productive forces … ?' The goal of human development is the 'development of the rich individuality which is as all-sided in its production as in its consumption'. Thus, the growth of human wealth is 'the absolute working-out of his creative potentialities', the 'development of all human powers as such the end in itself'. Within capitalism, however, the goal of capital is definitely not the development of that potential. Rather, as Marx wrote in *Capital* (1977, p. 772), the worker exists to satisfy the capitalist's need to increase the value of his capital 'as opposed to the inverse situation in which objective wealth is there to satisfy the worker's own need for development'.

In the co-operative society based upon common ownership of the means of production that Marx envisioned, the all-sided development of people would be based upon 'the subordination of their communal, social productivity as their social wealth', and their productive activity would flow from a unity and solidarity based upon recognition of their differences. Thus, the human community would be presupposed as the basis of production, and characteristic of this relation of associated producers would be that they expend 'their many different forms of labour power in full self-awareness as one single social labour force' (1977, p. 171). As a result of the focus upon human beings, increased productivity would come not at the expense of workers but would translate into the greater satisfaction of needs and free time – which 'corresponds to the artistic, scientific, etc. development of the individuals in the time set free, and with the means created, for all of them'. It would be 'time for the full development of the individual, which in turn reacts back on the productive power of labour itself as itself the greatest productive power'. All the springs of co-operative wealth would flow more abundantly, and the real products of this society of freely associated producers would be human beings able to develop their full potential in a human society.

THE PRODUCTS OF CAPITAL

But how do you get there? Some people look forward to the crises of capitalism in the expectation that their onset will lead workers to rise up in all their glory. They should read Marx more closely. No one was more aware than Marx of capital's strength and its domination of workers.

Marx understood quite well that the very nature of the wage–labour relation within capitalism produces workers who are conscious of their dependence upon capital. Having yielded to capital his *'creative power,* like Esau his birthright for a mess of pottage', the worker looks upon capital as 'a very mystical being' because it appears as the source of all productivity. 'All the productive forces of social labour appear attributable to it, and not to labour as such, as a power springing forth from its own womb.' Indeed, as Marx commented, the transposition of 'the social productivity of labour into the material attributes of capital is so firmly entrenched in people's minds that the advantages of machinery, the use of science, invention, etc. are *necessarily* conceived in this *alienated* form, so that all these things are deemed to be the *attributes of capital'* (Marx 1977, p. 1058). Wage labour assigns its own attributes to capital in its mind because the very nature of the capital–wage labour relation is one in which it does so in reality.

So, it is no accident at all that capital appears as the source of social productivity or that it looks as if the worker's well-being depends upon capital. Within the capital–wage labour relation, the worker needs capital. The very process of capitalist production produces and reproduces the working class that capital requires, workers who consider the necessity for capital to be self-evident:

> The advance of capitalist production develops a working class which by education, tradition and habit looks upon the requirements of that mode as self-evident natural laws. The organisation of the capitalist process of production, once it is fully developed, breaks down all resistance. (Marx 1977, p. 899)

Breaks down all resistance. Given Marx's statement that 'the great beauty of capitalist production' consists in its ability constantly to replenish the reserve army of labour and thereby to reinforce 'the social dependence of the worker on the capitalist, which is indispensable', how can we possibly talk about transcending capitalism (Marx 1977, p. 935)? On the contrary, as Marx noted about developed capitalism:

> In the ordinary run of things, the worker can be left to the 'natural laws of production', i.e. it is possible to rely on his

dependence on capital, which springs from the conditions of production themselves, and is guaranteed in perpetuity by them. (Marx 1977, p. 899)

THE PATH BEYOND CAPITAL

So, what did Marx propose in order to bridge the enormous gap between, on the one hand, workers who look upon the rule of capital as common sense and, on the other, a society in which associated producers recognise each other as differing limbs of a collective worker? For one thing, Marx dedicated himself to demonstrating to workers through his theoretical work that capital was simply the workers' own product turned against them. Precisely because of the inherent mystification of capital, Marx was determined to *demystify* capital, to reveal it to be not an independent source of productivity but, rather, the result of exploitation. Marx clearly believed that it was essential that workers grasp that it was the social productivity of the collective worker rather than capital which should be celebrated, and he considered this important enough to 'sacrifice my health, happiness, and family' to this end.

In addition to his theoretical work, however, Marx also was the central figure in the Working Men's International Association, an organisation that attempted to bring international unity to workers in their struggles against capital. The separation among workers benefits capital. Referring to the antagonism between Irish and English workers, he argued: 'It is the secret by which the capitalist class maintains its power. And that class is fully aware of it.' These two elements – challenging the ideological hegemony of capital by revealing its nature and working for unity in action against capital – were the ingredients in Marx's own recipe for transcending capitalism. Workers may be numerous, he noted in the Inaugural Address of the International, but they can succeed only 'if united by combination and led by knowledge'.

All of this presumed that workers were already engaged in struggle. Marx understood that workers in general engage regularly in warfare against capital, sometimes hidden, sometimes open. Not only because they are defending themselves against capital's drive for profits at their expense but also because 'the worker's own need for development' necessarily pits them against capital. Thus, the attempt of workers to satisfy socially generated needs for commodities leads to wage struggles, and their desire for time and energy for themselves underlies their struggles over the length and intensity of the workday. There are inherent limits, however, to how much success is possible from this guerrilla warfare conducted by groups of workers united in trade unions – given the power of capital as a whole, a power which rests upon its ownership of the products of labour.

Accordingly, Marx stressed the necessity for further unity as a class. Look at the Ten Hours' Bill, 'the marvellous success of this working men's measure', this victory for 'the political economy of the working class', he declared. Here was a case where, to limit the workday, workers in England recognised that they had 'to put their heads together and, as a class, compel the passing of a law' (Marx 1977, p. 416). Where workers do organise to pressure the state in this way, he commented, 'the working class do not fortify governmental power. On the contrary, they transform that power, now used against them, into their own agency.' Indeed, Marx argued, they could not succeed in limiting the workday otherwise: 'This very necessity of *general political action* affords the proof that in its merely economic action capital is the stronger side.'

Accordingly, it followed that workers should organise and struggle to gain political supremacy. This was the message of the International: 'To conquer political power has therefore become the great duty of the working classes.' And, this was the message that Marx and Engels continued to stress in the *Communist Manifesto*: 'the first step in the revolution by the working class is to raise the proletariat to the position of ruling class, to win the battle of democracy'.

But, then what? The *Manifesto* called for a process of making 'despotic inroads on the rights of property', introducing measures 'which appear economically insufficient and untenable, but which, in the course of the movement, outstrip themselves'. While the initial steps would not be far-reaching (the list of measures excluded nationalisation of private industry), the goal would be clear: 'The proletariat will use its political supremacy to wrest, by degrees, all capital from the bourgeoisie, to centralise all instruments of production in the hands of the State, i.e., of the proletariat organised as the ruling class; and to increase the total of productive forces as rapidly as possible.' The key term here is 'by degrees'. As Engels indicated in a second draft of the *Manifesto* (the 'Principles of Communism'), 'to abolish private property at one stroke' is no more possible than it is to increase productive forces to the necessary level at one stroke. The proletarian revolution 'will transform existing society only gradually'.

Thus, the picture that emerges from a consideration of the *Manifesto* is one involving a lengthy period during which the workers' state creates the foundations for a communist society. The state here is used to restrict the possibilities for the reproduction of capitalist property relations – while at the same time fostering the conditions for the emergence of state-owned property. And, the clear sense is that the process will be self-reinforcing. One measure will always lead on to the next, and 'the proletariat will see itself compelled to go always further'. It is a process, too, that would continue until such time that productive forces had been sufficiently developed to allow the last remnants of the old social relations to fall away. This political rule (or 'dictatorship') of the proletariat would exist within capitalism, and its success was the precondition for transcending capitalism.

An interesting picture of the organic development of the new society. Yet, there are two problems related to this scenario – one on the side of capital and the other on the side of wage labour. As noted, this process of despotic inroads upon capital is a path of gradualism. However, as Oskar Lange argued (Lange 1938, pp. 121–9), gradualism as a strategy ignores the response of capital to encroachments. If capitalists know in advance that the plan of the workers' state is to 'wrest, by degrees, all capital from the bourgeoisie', then their reaction will be predictable – no investment. The result will be crisis. Capital's response to 'despotic inroads' (and even minor ones) is to go on strike – which, given the ease with which capital moves in the modern capitalist world, can occur far more rapidly today than Marx could have anticipated. When capital goes on strike, the workers' state has two choices – give in or move in. Thus, Lange (p. 129) commented that, if a socialist government wants to do more than the administration of a capitalist economy, the only policy that an economist can recommend is 'a policy of *revolutionary courage*'.

This brings us, however, to a more serious concern about the *Manifesto*'s scenario. When we recall Marx's insights into the way capital produces workers who 'by education, tradition and habit' view its requirements as common sense, why assume that any of the struggles of workers are struggles to go beyond capital? Although workers struggle for higher wages and to reduce the length and intensity of the workday, why should we see this as any other than an attempt to get what they consider 'fair' for themselves within capitalism? Indeed, where workers attempt to use the state as 'their own agency' within capitalism (even where they have won 'the battle of democracy'), that state will be constrained to facilitate conditions for the expansion of capital – *as long as workers continue to view capital as productive.*

In short, as long as workers consider capital's requirements to be 'self-evident natural laws' and continue to function within the bounds of a relation in which the reproduction of wage labour as such requires the reproduction of capital, the response to crises – whether they are the result of economic forces or encroachments upon capital – will be to 'give in' rather than move in. Here, in a nutshell, is the sorry history of social democracy, which never ceases to reinforce the capital relation.

How can we reconcile, then, *Capital's* understanding of the 'social dependence of the worker upon capital' and the *Manifesto's* revolutionary programme?

REVOLUTIONARY PRACTICE

It is essential to understand that *Capital* is only about the logic of capital.[2] That was its point – to reveal the nature of capital and its tendencies. There is no place in *Capital* for living, changing, striving, enjoying, struggling and developing human beings. People who produce themselves through their own activities, who change their nature as they produce, are not the subjects of *Capital*. But, they *are* at the very core of Marx's understanding of the subjects of change. Human beings are beings of praxis – they are what they do, and when they engage in struggle, they transform themselves.

This is what Marx designated in his 'Third Thesis on Feuerbach' as 'revolutionary practice' – 'the coincidence of the changing of circumstances and human activity or self-change'. Marx's message to workers, as he would note subsequently, was that you have to go through years of struggle 'not only in order to bring about a change in society but also to change yourselves'. Over 20 years later, too, he wrote that workers know that 'they will have to pass through long struggles, through a series of historic processes, transforming circumstances and men'. Only in motion could people rid themselves of 'all the muck of ages and become fitted to found society anew'.

Thus, when workers struggle for higher wages, struggle against capital in the workplace and struggle for the satisfaction of their social needs in general, that very process is one of transforming them into people with a new conception of themselves – as subjects capable of altering their world. And, the same is true of the struggle to make the state the workers' agency. Not only is this necessary (because 'in its merely economic action capital is the stronger side'), but it is also an essential part of the process by which workers transcend their local interests and take shape as a class against capital as a whole. Thus, for example, the struggle to make the state expand its provision of use values *'needed for common satisfaction of needs* such as schools, health services, etc.' both attempts to substitute the state for capital as a mediator for workers and also unifies workers (skilled and unskilled, waged and unwaged). In this respect, the struggle for the state is an essential moment in the process of producing the working class as a class for itself, an essential moment in the process of going beyond capital.

But, what kind of state? It is essential to understand that Marx learned from workers – and never more so than with respect to the character of the state which workers need to serve them. Following the Paris Commune of 1871, Marx wrote that the Commune, the government initiated by the workers of Paris, proved that 'the working class can not simply lay hold of the ready-made state machinery, and wield it for its own purposes'; its particular character was 'the political form at last discovered under which to work out the economical emancipation of Labour'. (At last discovered!) In a word, the commune-form was the form of the 'dictatorship of the proletariat' described in the *Manifesto* – its purpose was 'to serve as a lever for uprooting the economical foundations upon which rests the existence of classes, and therefore of class-rule' (Marx and Engels 1971, pp. 68, 75).

The working class, Marx argued, could not use the existing type of state because it was infected – its very institutions involve a 'systematic and hierarchic division of labour', and it assumes the character of 'a public force organised for social enslavement, of an engine of class despotism' (pp. 68–9). How could the working class use such a state for its own purposes – a state whose very nature was hierarchy and power over all from above? Rather than being controlled *by* workers, such a state would represent the control *of* workers, retaining the character of a 'public force organised for social enslavement'. That is why Marx stressed that the commune was a 'Revolution against the *State* itself, of this supernaturalist abortion of society, a resumption by the people for the people of its own social life'. It was 'the reabsorption of the state power by society as its own living forces instead of as forces controlling and subduing it, by the popular masses themselves, forming their own force instead of the organised force of their suppression – the political form of their social emancipation' (pp. 152–3).

What, then, was the particular form of rule at last discovered? Simply, it was a decentralised government composed of councillors paid workers' wages and who were recallable and bound by the instructions of their constituents; in every district, common affairs would be administered by an assembly of delegates, and these assemblies would select deputies to constitute a central government. 'All France', Marx commented, 'would have been organised into self-working and self-governing communes' (pp. 155–6). This was the destruction of state power insofar as that state stood above society – 'its legitimate functions were to be wrested from an authority usurping pre-eminence over society itself, and restored to the responsible agents of society'. Centralised government would give way to the 'self-government of the producers' (pp. 72–3). And, yes, Marx responded to Bakunin's doubts (in notes on the latter's *Statehood and Anarchy*): all members of society *would* really be members of government 'because the thing begins with self-government of the township'.

In short, as Marx came to understand, we cannot be indifferent to the *form* of the state as an agency of workers. The form and the content of the state are inseparable. Only insofar as the state is converted 'from an organ superimposed upon society into one completely subordinate to it' can self-governing producers change both circumstances and themselves. As Marx knew, this new form of the state did not do away with class struggles. Indeed, it brings the producers together in their 'self-working and self-governing' assemblies and councils and calls upon them to drive beyond every barrier that capital puts up. What this struggle produces is an increasingly self-conscious collective worker, composed of workers who *cease* to be dependent upon capital – who are empowered by a state which is that of 'the popular masses themselves, forming their own force instead of the organised force of their suppression'.

When we consider the side of capitalism not developed by *Capital*, then we can reconcile the *Manifesto*'s revolutionary programme with Marx's analysis of the logic of capital. What Marx demonstrated in *Capital* was that the necessary condition for the reproduction of capital is the reproduction of the worker as wage labourer. Capital's 'indispensable' condition of existence is the feeling of dependence of the worker upon capital, and its inherent tendency is to produce that dependency. On the side of workers, on the other hand, we see an alternative logic driven by the workers' need for self-development that leads beyond that necessary condition of existence for capital. Even the more rapid onset of crises as the result of the mobility of capital would only compress the process of transcending capital. In this scenario, every step in the struggle against capital creates the basis for a deeper social relation among producers, one in which they emerge as a collective worker increasingly conscious of the interdependence of its limbs. And, it is this collective worker, working through its own state 'to wrest, by degrees, all capital from the bourgeoisie' which would be the premise for the 'cooperative society based upon the common ownership of the means of production'.

A BETTER WORLD IS POSSIBLE

But that second scenario, the workers' scenario, is not inevitable. And Marx knew this. That's why he insisted in 1853, that 'the continual conflicts between masters and men ... are ... the indispensable means of holding up the spirit of the labouring classes ... and of preventing them from becoming apathetic, thoughtless, more or less well-fed instruments of production'. Indeed, without strikes and constant struggle, the working classes 'would be a heart-broken, a weak-minded, a worn-out, unresisting mass'. It is the reason, too, why Marx was uncompromising in his criticism of all those who would 'dilute' class struggle, who would demobilise workers and put an end to 'proletarian snap'. Writing in 1879 against three such writers, he declared:

> For almost forty years we have stressed the class struggle as the immediate driving power of history and in particular the class struggle between bourgeoisie and proletariat as the great lever of the modern social revolution; it is, therefore, impossible for us to co-operate with people who wish to expunge this class struggle from the movement.

Why would any of this matter if the victory of workers is already recorded in the Good Book? Why, too, would he sacrifice 'health, happiness and family' to the writing of Volume 1 of *Capital* if it were all inevitable?

In the years since Marx wrote, that lack of inevitability is something we understand so much more clearly. The elements that concerned Marx are there even more strongly than before – the overwhelming mystification of the nature of capital and the separation and competition of workers internationally. But there is more now. Working people have recorded the failures of social democratic governments (which have demobilised and disarmed workers' movements and surrendered to capital) and 'actually existing socialism' (AES) (unreal socialist episodes marked by hierarchy and power from above). Small wonder that the declaration of TINA (there is no alternative) has resonated so deeply.

It is clear that one essential element in Marx's recipe for transcending capitalism, the centrality of revolutionary practice, was not written down – not at least in any place as consolidated as *Capital*. Had this ingredient received the attention it deserves as an indispensable part of Marx's theory, a wider opposition to the state-forms of social democracy and AES as structures that prevent the self-development of producers might have existed. In any event, given those experiences (which are now part of the collective memory of workers), the side of the self-development of human beings through their activity must be restored to its rightful place.

But, there is something else that needs to be added. Given his belief that workers would develop the elements of the new society in the course of their struggles, Marx was reluctant to write recipes for future cooks. But, he wrote at a time when utopian visions were commonplace. Now, after the experience of the last century with AES, it is an absolutely essential political task to demonstrate how AES was not consistent with Marx's vision. Given the widespread sense that 'socialism doesn't work' (which may be preceded by 'It's a lovely idea, but ...'), the problems identified in socialism need to be shown to be specific to AES and the *feasibility* of an alternative vision demonstrated.[3] (Simply asserting that 'that wasn't socialism' is not very convincing.) Thus, the vision of the society of associated producers needs to be part of the recipe. But, that's not a recipe for the cooks of the future. It's for today's cooks.

In short, there is no inevitability but there is possibility. Revealing capital as the workers' own product turned against them, working for unity in struggle, stressing the centrality of revolutionary practice for the self-development of the collective worker and setting out the vision of a feasible alternative – all these are essential ingredients for the demonstration that A Better World Is Possible. Build It Now.

REFERENCES AND FURTHER READING

Lange, O. (1938) *On the Economic Theory of Socialism*, ed. Benjamin E. Lippincott. New York: McGraw-Hill, 1964.

Lebowitz, M.A. (1991) 'The Socialist Fetter: A Cautionary Tale', in R. Miliband, R. and L. Panitch (eds.) *The Socialist Register: Communist Regimes: The Aftermath*. London: Merlin.

Lebowitz, M.A. (1992) *Beyond Capital: Marx's Political Economy of the Working Class*. New York: St. Martin's Press.

Lebowitz, M.A. (2000) 'Kornai and the Vanguard Mode of Production', *Cambridge Journal of Economics* 24(3), pp. 377–92.

Marx, K. (1973) *Grundrisse*. New York: Vintage Books.

Marx, K. (1977) *Capital*, vol. 1. New York. Vintage Books.

Marx, K. and Engels, F. (1971) *On the Paris Commune*. Moscow: Progress Publishers.

NOTES

1. I have chosen to use many direct quotations from Marx in this chapter – not to send the reader in search of the source but to convey Marx's point in language more compelling and relevant than mine. Most of the quotations from Marx are drawn from Volume I of *Capital* (Marx 1977), the only volume of *Capital* that Marx completed, and from his rich notebooks of 1857–58 which have been published as the *Grundrisse* (Marx 1973). Except where otherwise noted, these quotations appear (with proper citation) in Lebowitz (1992), an expanded version of which is forthcoming from Palgrave Macmillan.

2. Lebowitz (1992) explores this theme, focusing upon the other side of capitalism – the side of workers.

3. Some examples of my own efforts in this regard can be found in Lebowitz (1991, 2000). The subject is also the theme of a book in progress, *Studies in the Development of Communism: The Socialist Economy and the Vanguard Mode of Production*.

19 Towards a Society of Free and Associated Individuals: Communism

Paresh Chattopadhyay

Communism as the representation of an ideal society is at least as old as Plato. However, as a doctrine, it evolved only with the modern working class in the early nineteenth century. In particular Karl Marx and Friedrich Engels (mainly Marx) made communism famous as the projection of a society that could arise logically from the contradictions of capitalist society itself as the outcome of a self-emancipatory revolution of the working class. In this chapter we try to offer a coherent portrait of the new society as envisaged by Marx (and Engels). The chapter is divided into five sections. The first section touches on the (pre-)conditions for the advent of the new society. Then the succeeding sections deal respectively with the new mode of production, ownership relations, exchange relations and the relations of distribution in communism.

(PRE-)CONDITIONS OF COMMUNISM

In his projection of the communist society succeeding capitalism Marx, it should be pointed out, drew on the writings of his immediate predecessors – such as Saint Simon, Charles Fourier and Robert Owen – all of whom envisaged a post-capitalist society without exploitation of human by human. However, these pre-Marxian socialists advanced their ideas of the future society during a period of the undeveloped state of the working class and its movement as well as the absence of the material conditions of its emancipation. Hence they sought through their personal inventive action to create these conditions. For Marx (and Engels) on the other hand, the 'theoretical conclusions of the communists are in no way based on the ideas and principles that have been invented or discovered by this or that would-be universal reformer. They merely express the actual relations springing from an existing class struggle, from a historical movement going on under our very eyes' (1970, pp. 46–7).[1] In the same way Marx stressed that the material

conditions of the rise of communism – the union of free individuals – are not given by nature; they are the product of history.[2] The
future society arises from the contradictions of the present society.
The 'working class', writes Marx, 'have no ready-made utopias to
introduce ... They have no ideals to realise, but to set free the
elements of the new society with which old collapsing bourgeois
society itself is pregnant' (in Marx and Engels 1971, p. 76). Marx
argues that at a certain stage of capitalism's development, its
relations of production – that is, relations in which individuals enter
in the social process of production of their lives – turn into fetters for
the further development of the forces of production – including the
'greatest productive force', the working-class forces which have been
engendered by capital itself and have progressed under it hitherto.
This signifies that the old (capitalist) society has reached the limits
of its development, and that it is time for it to yield place to a new,
higher social order – which thus signals the beginning of the 'epoch
of social revolution' (in Marx and Engels 1970, pp. 181–2).

On the other hand, before departing from the scene, capital,
besides engendering and developing the subjective agents of the
revolution – its 'grave diggers', the working class (the proletariat) –
will have already created the necessary material conditions for the
advent of the new society without which 'all attempts to explode
[current society] would be quixotic (Marx 1981, p. 159).[3] Put briefly,
these conditions are the great abundance of material wealth based on
the universal development of the productive forces and, necessarily
connected therewith, the socialisation of labour and production. It
cannot be sufficiently stressed that though capital creates the
material conditions of its own disappearance as well as those for the
advent of the new society, the old society is not revolutionised
within itself on its own simply because these material conditions
exist. It is the working class (the proletariat), capital's 'wage slaves',
which is the active agent for eliminating capital and building the
communist society. And it is only through a communist revolution
that the working class can be emancipated, ceasing to be wage
labourers and becoming 'associated labourers'. There are two points
to stress here. First, this working class or proletarian revolution is
self-emancipatory. Marx stresses that 'the emancipation of the
working classes must be conquered by the working classes
themselves' (*General Council Minutes* 1964, p. 288).[4] Secondly, the
(self-)emancipation of the proletariat signifies, at the same time, the
emancipation of the whole humanity, the proletariat being the

lowest class in capitalist society, which generates the 'last antagonistic form of the process of social production' (in Marx and Engels
1970, p. 182).

The communist revolution *begins* with the smashing of the
bourgeois political rule and the establishment of absolute political
rule by the proletariat – the rule of the immense majority in the
interest of the immense majority, the 'conquest of democracy' (Marx
and Engels 1970, p. 52, translation modified). This so-called 'seizure
of power' by the proletariat in no way signifies the *victory* of the
revolution which, on the contrary, continues through a more or less
prolonged 'transition period', a 'period of revolutionary transformation' of the capitalist society into the communist society, during
which the whole bourgeois mode of production and therewith the
whole bourgeois social order are superseded (Marx, in Marx and
Engels 1970, p. 327). Until capital completely disappears, the workers
do not cease to be proletarians, and hence the proletarian rule – the
'*revolutionary dictatorship of the proletariat*',[5] as Marx calls it (in Marx
and Engels 1970, p. 327, underlined in the original) – continues
throughout the 'transitional period', the period of preparation for
the workers' self-emancipation. At the end of the process, with the
disappearance of capital, wage labour also naturally disappears. The
proletariat with its political rule ceases to exist, leaving individuals
as simple producers; classes come to an end along with the state, the
embodiment of class domination and oppression.[6]

NEW MODE OF PRODUCTION

The outcome of the workers' self-emancipatory revolution is the
communist (socialist) society.[7] In all hitherto existing societies –
based on class rule – community has in fact stood as an independent
power against singular individuals and subjugated them. Thus it has
really been a 'false' or 'illusory' community. In the communist
society, in sharp contrast, there arises, for the first time, the 'true'
community where universally developed individuals dominate their
own social relations (Marx and Engels 1974, p. 83; Marx 1975,
p. 265; 1981, pp. 162–4; 1978a, pp. 82–3). This is what Marx calls
'association' or 'union' of 'free individuals' based on a new mode of
production – the 'communist' or 'associated mode of production' –
in which the 'free development of each is the condition of the free
development of all' (Marx and Engels 1970, pp. 49, 53; Marx 1978a,
p. 82; 1984b, pp. 440, 607). The term 'union' or 'association' in this
connection has a profound meaning.

Under capitalism individuals in society do not relate to one another directly. Relations between individuals take the form of relations between their products, which appear as commodities to be exchanged, take on an autonomous existence vis-à-vis the producers and dominate them. Thus individuals are 'alienated' from one another as well as from their own products. Additionally, producers separated from the means of production (within the broad 'conditions of production', their own creation) have, in order to survive, only their labour power to fall back on, which they are compelled to sell to the owners of the means of production against a wage or salary. Thus 'union' or 'association' as applied to the communist society has a double meaning as opposed to capitalism's 'alienation' and 'separation'. First, it is a voluntary and unmediated union of individuals dominating their own products, which are no longer commodities, and secondly, it is an unconstrained union of producers with the conditions of production (which puts an end to the producers' situation as wage labourers or, as Marx calls them, 'wage slaves').

This union of producers with the conditions of production, opposed though it is to capitalism's alienation and separation, is not, it should be stressed, a simple restitution of the type of union which prevailed under pre-capitalism – such as constrained union under slavery or serfdom or voluntary union under simple family enterprise or 'natural communism'. Under them neither could there be a universal development of the productive powers of labour – engendering an abundance of material wealth – nor could labour and production be socialised at a universal level. As referred to earlier, it is only capital that by separating the producers from the conditions of production and pursuing the path of production for production's sake – the logic of accumulation – creates, independently of the individual capitalist's will, these fundamental material conditions for building the new society (Marx 1978c, pp. 422–3). Marx envisages the whole process of human development in terms of the changing relation between the 'Man of Labour' and the 'Instruments of Labour', starting with their *Original Union*, then passing on to their *Separation* and finally arriving, through a 'new and fundamental revolution in the mode of production', at the 'restoration of the original union in a new historical form' (in 1970, p. 208, capitalisation and emphasis in original).

Individuals in the new society are free in a sense unknown hitherto. Going beyond the 'personal dependence' (domination of person over person) of pre-capitalism, as well as the 'material dependence' (domination of the product over the producer) of capitalism, 'social individuals' attain their 'free individuality' in this 'society of free and associated producers'. It is, as Marx calls it, the 'real appropriation of the human essence by and for the human', a 'complete elaboration of the human interiority', 'the development of human energy which is an end in itself' (1975, p. 348; 1981, p. 488; 1984b, p. 820; translation modified).[8]

RELATIONS OF OWNERSHIP

The social relations of production form the 'real basis' of a society. The latter's ownership relations – relations around the ownership of the means of production – arise from and are simply the juridical representation of the (social) relations of production (Marx, in Marx and Engels 1970, pp. 181, 318). Hence when the latter relations are transformed, society's ownership relations are also transformed. In all class societies (including capitalist society) the great majority of labouring individuals has been deprived of ownership of the means of production, which have been owned by only a small minority. Private ownership in this fundamental *class* sense, never recognised by jurisprudence, has prevailed till now. Marx calls it 'private ownership of a part of society'. Under capitalism it signifies capitalist-class ownership of the means of production, which is only the reverse side of the 'non-ownership' or 'alien property' of the means of production for the labourers (1978b, p. 56; 1978c, p. 460; 1976, p. 1003). This is independent of the question of ownership by *individual* capitalists in their private capacity.[9] Within this broad class ownership there could be different forms of private ownership. In bourgeois jurisprudence (taken over from the Roman law), as well as in the commonly accepted sense, private ownership refers to the ownership (of the means of production) by an individual (a household) or by a business enterprise. Quite logically the juridical replacement of this form of ownership by 'public' (basically state) ownership has meant the abolition of capitalist private ownership as such in the means of production.[10] However, this is a mistaken view. It confuses the ownership *form* with the ownership *relation* itself which is simply the juridical representation of the production relation of a society. The capitalist (class) ownership relation is given as soon as the capitalist production relation is given. This specific ownership relation is defined by the producers' *separation* from the means of production – they themselves being neither owners nor forming part of the means of production. This ownership *relation*, however, could assume different *forms* such as ownership of the 'individual capitalist' or the 'associated capitalist' (joint stock companies) or even by the 'state' (Marx 1984a, pp. 100, 237). Marx in fact shows that capitalist-class ownership has taken different forms following the demands of capital accumulation. Originally starting as pure private ownership by individual capitalists, capitalist ownership reaches a stage where private ownership in its initial form is eliminated, the means of production come under collective capitalist ownership (as in share companies) and capital becomes 'directly social' as distinct from 'private capital' which, given the continuation of the capitalist relation of production, Marx designates as the 'abolition of capital as private ownership within the framework of capitalist production itself' (1984b, p. 436). Marx (1978a, p. 588) even envisages a stage where, driven by the needs of accumulation, the capital of the whole society comes under single ownership (though capital as a production or ownership *relation* is not eliminated thereby).

Thus state ownership of the means of production does not at all mean the abolition of capitalist-class ownership – that is, capitalist private ownership in the sense argued earlier – so long as the producers remain separated from the means of production and continue to be wage labourers. It simply signifies the end of the legally recognised private *individual* (including corporate) ownership of the means of production. Here the 'real agents of capitalist production' or the 'functionaries of capital', as Marx would call them, are at the same time the functionaries of the state.

Indeed the *Communist Manifesto* underlines the need for the juridical elimination of *individual* private ownership in the means of production and for bringing it under the ownership of the proletarian political power only as a *beginning* measure. Since the setting up of the workers' political power does not mean the immediate supersession of the capitalist relations of production (the inauguration of communism), the proletarian state ownership does not at all mean the end of capitalist private ownership in the fundamental (class) sense. In this latter sense the 'abolition of private ownership' is equated in the *Manifesto* only with the 'disappearance of class ownership' (Marx and Engels 1970, pp. 47, 49, 52). Exactly the same idea Marx expresses many years later in his discourse on the Paris Commune (in Marx and Engels 1971, p. 75). Capitalist-class ownership disappears only with the disappearance of capitalist production relations. Capitalist private ownership – both in its individual sense and in its fundamental class sense – yields place to collective (social) appropriation of the conditions of production under communism. This is in fact the appropriation of the conditions of production by society itself, which is only the collectivity of the free producers who are now 'universally developed social individuals'. (The disappearance of the state with the disappearance of capital was noted earlier.) This appropriation, contrary to its earlier forms, which had a limited character (involving private class ownership), has now a total and universal character. This is because the deprivation of the lowest class – the 'immense majority', as the *Manifesto* puts it – within the last antagonistic social formation is total; and secondly, given the universal character of the development of the productive forces attained under capitalism, the appropriation of the forces of production has also to be universal, appropriation by society (of emancipated producers) itself as an entity. Thereby the individual in the new society also becomes a 'total' or 'integral' individual. Quite appropriately Marx calls this transformed ownership, arising from the 'expropriation of the expropriators', 'individual ownership' (1975, p. 348; 1974, p. 93; in Marx and Engels 1971, p. 75; 1978a, p. 715).

EXCHANGE RELATIONS

With the transformation of society's production relations, its exchange relations – that is, individuals' material exchanges with nature and their social exchanges among themselves – are also transformed. Capital, while increasing at an unprecedented scale the material forces of production, rendering humans less dependent on nature's caprices, at the same time, driven by the logic of accumulation, seriously damages the environment and undermines the natural powers of the earth together with those of the human producer, the 'twin fountains of all wealth' (Marx 1978a, pp. 474–5; 1984b, p. 813). In the new society, freed from the mad drive for accumulation and with the unique goal of satisfying human needs, individuals rationally regulate their material exchanges with nature with the least expenditure of force and carry on these exchanges in the conditions most worthy of and in fullest conformity with their human nature (Marx 1984b, p. 820).

Coming to the exchange relations among individuals, first let us note that in any society the labour of the individual producers creating useful objects for one another has, by that very fact, a social character. However, in a commodity (capitalist) society, where products result from private labours executed in reciprocal independence, the social character of these labours – hence the relations of the creators of these products – are not established directly (see Chapter 1). Their social character is mediated by exchange of products as commodities. As mentioned earlier, the social relations of individuals take the form of social relations of their products. The products dominate the producers confronting them as an independent power. The related labour is 'alienated labour' (Marx 1975, pp. 330–1). The whole process of this 'mystification' Marx famously calls in *Capital* 'commodity fetishism'.

With the inauguration of the 'union of free individuals', there begins, as noted above, the collective appropriation of the conditions of production by society. Consequently, with the end of private appropriation of the conditions of production, there also ends the need for the products of individual labour to go through exchange, taking the commodity form. In the new society individual labour is *directly social* from the beginning. In place of exchange of products taking commodity form (as in the old society) there is now 'free exchange of activities' among 'social individuals' determined by their collective needs and aims on the basis of collective appropriation. In the communist society, in contrast with the capitalist society, the social character of production is *presupposed*, and participation in the world of products is not mediated by the exchange of reciprocally independent labours or of products of labour. Here the labour of the individual is posited as social labour from the very outset. The product of the individual 'is *not an exchange value*' (Marx 1981, p. 172). In a famous text Marx asserts that in the 'communist society as it has just *come out* of the capitalist society the producers do not exchange their products and as little does labour on these products appear as value' (in Marx and Engels 1970, p. 319, emphasis in text).[11]

ALLOCATION AND DISTRIBUTION

The basic distribution in any society is the distribution of the conditions of production – the material means of production as well as the living labour power – from which follows the distribution of the products of these conditions. The 'distribution of the conditions of production is a character of the mode of production itself'. Hence the mode of distribution based on the capitalist mode of production is transformed with the transformation of this mode into the associated mode of production. For any society, the distribution of the conditions of production is really the allocation of the total labour time (dead and living) of society across the economy in definite proportions corresponding to its needs. Equally, society's total time employed on production (and related activities) has to be economised in order to leave the maximum non-labour time for the enjoyment and self-development of society's members. Thus 'all economy is finally reduced to the economy of time'. The economy of time as well as its distribution among the different branches of economic activity are executed in different societies in different ways. Whereas under capitalism the distribution of society's labour time is effected through exchange of products taking commodity form, under the 'Republic of Labour' the problem is solved through the direct and conscious control of society over its labour time, without the need for social relations of persons to appear as relations between things (Marx's letters to Engels and Kugelmann, 8 January and 11 July 1868). On the other hand, given the unusual importance that the communist society would attach to the self-development and enjoyment of 'social individuals', requiring maximum non-labour time beyond the labour time necessary to satisfy their material needs, the economy of time, including its distribution, would become the 'first economic law on the basis of communal production' (Marx 1981, p. 173).

The economy of society's global time employed in material production (and related activities) – generating disposable time thereby – acquires a new meaning in the new society. This surplus labour time beyond the time required for labourers' material needs, instead of being appropriated by a small minority in the name of society (as in all class societies) now becomes *society's* free time creating the basis of all-round development of the 'associated producers'. In fact the distinction between necessary and surplus labour time loses its earlier meaning. Necessary labour time would now be measured in terms of the needs of the 'social individuals' and not in terms of the needs of valorisation. Surplus labour time – now the free time for the 'associated producers' – would mean free activity which, unlike labour time as usually understood, would not be determined by any external finality that has to be satisfied either as a natural necessity or as a social obligation (Marx 1978c, p. 257).

Turning to the distribution of the total social product in the new society, it is first of all divided between the production needs and the consumption needs of society. As regards the share going towards production needs, it is again divided between the replacement and extension of society's productive apparatus on the one hand and society's insurance and reserve funds against uncertainty on the other. The rest of the social product serves collective consumption – such as health, education, provision for those unable to work – and personal consumption. As regards the mode of distribution of the means of consumption – that is, after deducting from the common fund – among society's producing individuals, the latter, collectively mastering and appropriating the conditions of production and hence ceasing to be sellers of labour power, no longer receive the return to their labour in wage form. Instead they receive from their (own) 'association' a kind of labour certificate indicating the amount of labour each one has contributed to production, enabling the person to draw from the common stock of means of consumption an amount costing the same amount of labour. These certificates are *not* (in the absence of commodity production) money; they do not circulate (Marx 1978, pp. 97–8; 1984a, p. 362; in Marx and Engels 1970, p. 319).

At the initial stage of the new society – just coming out of the womb of capital – this principle of equivalent exchange (labour against labour of the same amount), similar to but *not the same* as what prevails under commodity production, cannot be avoided. This process is wholly overcome only at a higher stage of the 'association' when the opposition between physical and mental labour vanishes, when labour becomes life's prime need and not simply a means of living, when all-round development of the 'social individual' along with the development of the productive forces takes place and when all the springs of 'co-operative wealth' flow more fully. Only then would prevail the principle: 'from each according to one's ability, to each according to one's needs' (Marx, in Marx and Engels 1970, p. 320).

REFERENCES AND FURTHER READING

Engels, F. (1970) 'Socialism: Utopian and Scientific', in K. Marx and F. Engels, *Selected Works in One Volume*. Moscow: Progress Publishers.

General Council of the First International 1864–1866: Minutes (1964). Moscow: Progress Publishers.

Marcuse, H. (1960) *Reason and Revolution*, part 2. Boston: Beacon Press.

Marx, K. (1975) 'Economic and Philosophical Manuscripts', in *Early Writings*. New York: Vintage Books.

Marx, K. (1976) 'Results of the Immediate Process of Production', in *Capital*, vol. 1. Harmondsworth: Penguin.

Marx, K. (1977) *Selected Writings*, ed. D. McLellan. Oxford: Oxford University Press.

Marx, K. (1978a, 1984a, 1984b) *Capital*, vols. 1, 2, 3. Moscow: Progress Publishers.

Marx, K. (1978b, 1978c) *Theories of Surplus Value*, vols. 1, 3. Moscow: Progress Publishers.

Marx, K. (1981) *Grundrisse*. Harmondsworth: Penguin.

Marx, K. and Engels, F. (1970) *Selected Works in One Volume*. Moscow: Progress Publishers; including the *Communist Manifesto* (pp. 31–63), Marx's Preface to the 'Contribution to the Critique of Political Economy' (pp. 180–4), 'Wages, Price and Profit' (pp. 185–226) and 'Critique of the Gotha Programme' (pp. 315–31).

Marx, K. and Engels, F. (1971) *On the Paris Commune*. Moscow: Progress Publishers.

Marx, K. and Engels, F. (1974) *The German Ideology*, part 1. London: Lawrence & Wishart.

Rosdolsky, R. (1977) *The Making of Marx's Capital*, part 6, ch. 28. London: Pluto Press.

NOTES

1. 'Communism for us is not a *state* (of things), which should be created, an *ideal* to which the reality should conform. We call communism the *real* movement which abolishes the existing state (of things). The conditions of this movement result from the premises existing at present' (Marx and Engels 1974, pp. 56–7, translation modified).

2. The young Marx already wrote: 'In order to supersede the *idea* of private property the *idea* of communism is totally sufficient. In order to supersede private property as it really exists, *real* communist activity is necessary. History will give rise to such activity, and the movement that we already know in thought to be a self-superseding movement will in reality undergo a very difficult and protracted process' (1975, p. 365, original emphasis, translation slightly modified).

3. 'No social formation ever perishes before all the productive forces, for which it is large enough, have developed, and new, higher relations of production never appear before the material conditions of their existence have been hatched within the womb of the old society itself' (Marx, in Marx and Engels 1970, p. 182, translation slightly modified).

4. Contrary to a fairly widespread idea of the Left, the workers themselves through their own experience of struggle against capital, unaided by any outside 'guide', arrive at the consciousness of the necessity of revolution to free themselves from capital's subjugation. As Marx and Engels underline, 'The consciousness of the necessity of a profound revolution arises from the (working) class itself' (1974, p. 94).

5. Referring to the Paris Commune under workers' rule (1871), Marx observes that the 'superseding of the economical conditions of the slavery of labour by the conditions of free and associated labour can only be the progressive work of time ... in a long process of development of new conditions ... through long struggles, through a series of historic

processes transforming circumstances and men' (in Marx and Engels 1971, pp. 76, 156–7).

6. In 1847, in the *Poverty of Philosophy*, Marx wrote that in course of its development the 'labouring class will substitute for the civil society an association which will exclude classes and their antagonism and there will no longer be political power in the proper sense' (Marx 1977, p. 215). The very next year he (and Engels) stressed that 'in the course of development (of the revolution) all production is concentrated in the hands of the associated individuals (and) public power loses its political character' (1970, p. 53; translation modified). Years later Marx praised the Parisian communards for their 'revolution against the *state* itself, not against this or that form of state power' (in Marx and Engels 1971, p. 152; original emphasis).

7. Marx does not distinguish between communism and socialism. Both stand for the society succeeding capitalism. The distinction was first to be made famous, if not introduced, by Lenin. Marx, of course, distinguishes between a lower and a higher phase of the new society – that is, of the same post-capitalist society – depending on the stage of development (in Marx and Engels 1970, pp. 320–1). However, starting with Lenin, the Left has, by and large, misleadingly treated these two phases as two distinct successive societies – 'socialism' and 'communism'.

8. Indeed, corresponding to the three stages of evolution of the relation between the producers and the conditions of production, just mentioned, Marx mentions these three stages of the development of human freedom – the 'personal dependence' of pre-capitalism where human productivity develops only in small proportions and at isolated points, personal independence based on the 'objective dependence' of capitalism – the 'second great form' in which alone is formed a system of general social metabolism, universal relation, all-sided needs and universal faculties – and, finally, '[f]ree individuality based on the universal development of individuals and on their subordination of their collective, social productivity as their social wealth', the third form. The 'second', Marx adds, 'creates the conditions for the third' (1981, p. 158).

9. It should be stressed that Marx conceives the individual capitalist as a mere 'functionary of capital' the 'real agent of capitalist production', not necessarily a private owner of capital, receiving a 'wage of administration' for exploiting the labour of others in the 'real process of reproduction' (1978c, p. 477; 1984b, pp. 382–3, 387–8, 436).

10. The 'really (non-)existing socialist' societies have justified their 'socialism' essentially on the basis of this logic.

11. It is immediately clear that the so-called 'market socialism' – touted by a section of the Left as an alternative to capitalism – is a contradiction in terms (as we noted earlier, Marx does not distinguish between socialism and communism). Either you have the market as the basic exchange relation of society, in which case you have a capitalist society, or you have 'free exchange of activities' among individuals unmediated by the commodity form of the product of labour, in which case you have a communist or socialist society.

Contributors

Suzanne de Brunhoff is honorary research director, Centre National de la Recherche Scientifique, and taught economics at different French and foreign universities. She is the author of several books, including *Marx on Money* (New York: Urizen Press, 1976) and *The State, Capital and Economic Policy* (London: Pluto Press, 1976).

Paul Burkett teaches at the department of economics, Indiana State University, Terre Haute, Indiana. He is a long-time member of the Conference of Socialist Economists and the Union for Radical Political Economics, and the author of *Marx and Nature: A Red and Green Perspective* (New York: St. Martin's Press, 2000).

Paresh Chattopadhyay teaches political economy at the University of Québec in Montréal. He is the author of *The Marxian Concept of Capital and the Soviet Experience* (Westport, Conn.: Praeger, 1994).

Simon Clarke is professor of sociology at Warwick University, UK. Since 1990 he has been researching workers' organisation, labour and employment in Russia in collaboration with Russian and international trade union organisations. In addition to his work on Russia, his most recent books are *The State Debate* (ed., London: Macmillan, 1991) and *Marx's Theory of Crisis* (London: Macmillan, 1994).

Christopher Cramer teaches political economy at the School of Oriental and African Studies (SOAS), University of London. He has worked chiefly in Southern Africa, and in 2000 launched at SOAS the MSc in Violence, Conflict and Development.

Elizabeth Dore teaches Latin American history at the University of Southampton, UK. Her books include *Gender Politics in Latin America* (ed., New York: Monthly Review Press, 1997), *Hidden Histories of Gender and the State in Latin America* (ed., with M. Molyneux, Durham, N.C.: Duke University Press, 2000), *The Peruvian Mining Industry: Growth, Stagnation and Crisis* (Boulder, Colo.: Westview Press, 1988) and *The Myth of Modernity: Peonage and Patriarchy in Rural Nicaragua* (Durham, N.C.: Duke University Press, forthcoming).

Ben Fine is professor of economics at SOAS, University of London. He is the author of several books, including *Labour Market Theory: A Constructive Reassessment* (London: Routledge, 1998) and, most recently, *Social Capital versus Social Theory: Political Economy and Social Science at the Turn of the Millennium* (London: Routledge, 2001) and *The World of Consumption: The Material and Cultural Revisited* (London: Routledge, 2002).

Diego Guerrero is professor of applied economics at the Complutense University, Madrid. Since 1995 he has been co-ordinating the doctorate programme in his department, and researching on competitiveness, macroeconomics, globalisation and heterodox economic thought. His most recent books are *Manual de Economía Política* (ed.) (Madrid: Síntesis, 2002) and *Lecturas de Economía Política* (ed.) (Madrid: Síntesis, 2002).

John Holloway is professor at the department of sociology, Instituto de Ciencias Sociales y Humanidades, Benemérita Universidad Autónoma de Puebla, Puebla, Mexico. Has published widely on Marxist theory and the theory of the state. His publications include: *Change the World Without Taking Power* (London: Pluto Press, 2002), *Zapatista! Reinventing Revolution in Mexico* (ed., with Eloína Peláez, London: Pluto 1999), *Open Marxism III* (ed., with others, London: Pluto Press, 1995), and *Global Capital, National State and the Politics of Money* (ed., with W. Bonefeld, London: Macmillan, 1995).

Costas Lapavitsas teaches economics at SOAS, University of London. He works on finance, monetary theory, history of economic thought, and the Japanese economy. He is the author (with M. Itoh) of *Political Economy of Money and Finance* (London: Macmillan, 1999).

Michael A. Lebowitz teaches economics at Simon Fraser University, Vancouver, Canada. He has been an editor of *Studies in Political Economy: A Socialist Review* since 1980. In addition to preparing an expanded edition of his *Beyond Capital: Marx's Political Economy of the Working Class* (London: Macmillan, 1992), he is working on a book on the theory of socialist economies, *Studies in the Development of Communism: The Socialist Economy and the Vanguard Mode of Production* and on a collection of his essays on 'Analytical Marxism', methodology and crisis theory.

Les Levidow is a research fellow at the Open University, where he has been studying the safety regulation and innovation of agricultural biotechnology since 1989. This research encompasses the European Union, and the United States of America and their trade conflicts. He also has been managing editor of *Science as Culture* since its inception in 1987, and of its predecessor, the *Radical Science Journal*. He is co-editor of several books, including: *Science, Technology and the Labour Process*; *Anti-Racist Science Teaching*; and *Cyborg Worlds: The Military Information Society* (London: Free Association Books, 1983, 1987, 1989).

Simon Mohun teaches economics at Queen Mary, University of London. He has published widely on value theory, including *Debates in Value Theory* (ed., London: Macmillan, 1994), and his research focuses on the empirical application of Marxian value categories to contemporary capitalist economies.

Fred Moseley is professor of economics at Mount Holyoke College, South Hadley, Massachusetts. He is the author of numerous books and articles on Marxian value theory and crisis theory, including *The Falling Rate of Profit in the Postwar United States Economy* (London: Macmillan, 1991), *Marx's Method in Capital: A Reexamination* (Atlantic Highlands, N.J.: Humanities Press, 1993), and *Heterodox Economic Theories: True or False?* (Aldershot: Edward Elgar, 1995).

Michael Perelman teaches economics at California State University, Chico. His most recent books are *Steal this Idea: Intellectual Property Rights and the Corporate Confiscation of Creativity* (New York: Palgrave, 2002), *The Pathology of the US Economy Revisited: The Intractable Contradictions of Economic Policy* (New York: Palgrave, 2001), *The Invention of Capitalism: The Secret History of Primitive Accumulation* (Durham: Duke University Press, 2000), *Transcending the Economy: On the Potential of Passionate Labour and the Wastes of the Market* (New York: St. Martin's Press, 2000), *The Natural Instability of Markets: Expectations, Increasing Returns and the Collapse of Markets* (New York: St. Martin's Press, 1999) and *Class Warfare in the Information Age* (New York: St. Martin's Press, 1998).

Alfredo Saad-Filho teaches political economy of development at the department of development studies, SOAS, University of London.

He is the author of *The Value of Marx: Political Economy of Contemporary Capitalism* (London: Routledge, 2002).

John Weeks is professor of development economics at SOAS, University of London. He is the author of several books on development theory and Marxist analysis, including *Capital and Exploitation* (Princeton: Princeton University Press, 1981).

Ellen Meiksins Wood is the author of several books, including *The Retreat from Class* (London: Verso 1986, 1998), *The Pristine Culture of Capitalism* (London: Verso, 1991), *Democracy Against Capitalism* (Cambridge: Cambridge University Press, 1995), and *The Origin of Capitalism: A Longer View* (London: Verso, 2002).

Paul Zarembka is professor of economics at the State University of New York at Buffalo. He has been editor since 1977 of *Research in Political Economy* (JAI/Elsevier; web: ourworld.compuserve.com/homepages/PZarembka). His work on accumulation is represented by lengthy articles on 'The Accumulation of Capital, Its Definition', on 'Rosa Luxemburg's *Accumulation of Capital*', and on 'Accumulation of Capital in the Periphery', as well as a shorter piece on primitive accumulation. Other topics have included the Soviet system, Poland, and the declining importance of Hegel for Marx, and he is currently completing research on Lenin's economics. He has lived and worked in Switzerland, Germany, Poland and France, in addition to California and New York, and has been a union activist.

Index

Printed by Printforce, United Kingdom